If only you could see yourselves for even a
brief moment from our perspective, you would
re-member Home and your true heritage.
Perhaps then you would treat yourselves and
each other as the masters you truly are.

ESPAVO!

the Group

Re-member

A Handbook for Human Evolution

Re-member
A Handbook for Human Evolution

Published by: **Lightworker**

P.O. Box 1496
Poway Ca. 92074-1496 USA
www.Lightworker.com

The authors and publishers of this book do not dispense medical advice or prescribe any technique as a form of treatment for physical or emotional problems and therefore assume no responsibility for your actions. The intent of this material is to provide general information to help your quest for emotional and spiritual growth. We encourage you to seek professional assistance for all areas of healing.

Lightworker Books and Tapes can be purchased in retail stores or by telephone. Toll free (877) 248 5837
Outside the U.S. please call 1 (858) 748-5837

Cover and Book Design and Layout courtesy of Terrel Baker of Terrel Baker Creative, Golden, Colorado

Re-member – A Handbook for Human Evolution
Written by Steve Rother
Edited by Sandra Sedgbeer
Copyright ©2000 – Lightworker
Printed in the United States of America
First Edition - First Printing – October 2000   Second Printing - November 20

Lightworker   is a registered service mark with the United States Patent and Trademark Office

ISBN# 1-928806-08-2 $14.95

Acknowledgments

This book was an effort of many people who selflessly came forward to be a part of the presentation of this information. We are honored to present them here:

A special thanks to a talented and dear friend that we are very proud to call our editor: *Sandra Sedgbeer*

Phyllis Brooks
Gayle Schildt
David Solinger
Terrel Baker
Claire Gibb
Sharyl Jackson
Lorena Solinger
Carol Holaday
Eve Meng
Christine Szynal
Natalia Kavtaradze
Tami O'Sullivan
Joe Moriarity
Anne Sabin
Colette Halpin
Austin Rother
Catherine Kasper
Laura Placeras
Trude Kopecek
Raymond Murphy
Wendy Marie Collins
Andrea Post
Debbie Puerner
Deborah Velting
Emily Green
Eva Reinermann
Gary and Laura Grimshaw

Peter Hyman
Lee Carroll
Patty Gleeson
Len Delekta
Ka-Sandra Love
Morgan Kiilehua
Ted and Donna Dircz
Jan Wilson
Charlotte Ruhl
Rie Forsell
Lourdes Resperger
Greg Lambert
J.P. Dery
Penny Johnson Dare
Ingrid Kramer
Brent Rother
Cynthia Carr
Shala Mata
Mary Ellen Kelly
Nancy Mott
Heather Richmond
Caliesha Stewart
MaryLynn Schmidt
Mary K Smith
John Kubik
Bill Goforth
Brenda Pakkala

None of this would be possible without the love, support and encouragement of my partner in Love and Lightwork:

Barbara Redington Rother

Table of Contents

In Memory of Jack Rother. . .

A Lightworker who was slightly ahead of his time.

This is for you, Dad

An Unusual Note from the Editor

*I*t is not often that an editor feels compelled to add a note to an author's book. But by the time you are through reading this, I hope you will understand why I felt that, on this particular occasion, it was not only important to do so, but possibly even essential to your understanding and enjoyment of this book.

As someone who has spent their entire career working in the media as a journalist, magazine editor, publishing consultant and the author of six books of my own, I've naturally encountered many challenges over the years. In all honesty, however, I have to say that nothing in my career thus far has been as challenging and frustrating, or indeed as truly bizarre, as editing this book.

Given my background, it may not surprise you to learn that I am somewhat of a 'stickler' for clarity, punctuation and good grammar. Imagine my surprise, therefore, when, over and over again, I found that many of the text cuts, edits, corrections in grammar, punctuation and syntax, as well as several of my own attempts to clarify some of the information in the channels, kept disappearing from my computer. At first I dismissed this as a simple, stupid failure on my own part to hit the "save" button. The second time I began to wonder whether there was a technical malfunction. But when it kept on happening, often with the very same passages I had been 'clarifying', I started to wonder whether there might be another explanation…

When I voiced my concerns to Steve, he fell about laughing. For, of course, Steve understood something that I, in my enthusiasm to do a "proper, professional editing job", had failed to consider. The information in this book had come from the Group. Thus, what was more natural than that they should be hovering over my shoulder, approving - or not, as the case clearly proved often to be - the changes I was making?

To say that I was totally flabbergasted by their involvement is an understatement. After all, an editor's job is to guide his or her author, to help them clarify what is unclear, to tidy up their work by showing them where they may have repeated themselves, to edit extraneous passages... and so on and so forth. Fortunately, Steve was always very amenable and co-operative when it came to being 'guided' by his 'editor', but clearly the same could not be said of the Group.

The lesson I learned from this is that it is one thing to guide an earth-bound, in-the-flesh, human author, such as Steve. It is quite another thing to deal with a group of non-corporeal, higher entities whose sole purpose is to guide, rather than be guided. Steve says that the Group has a wonderful sense of humor. All I can say is, they must have been clutching their sides and falling off their heavenly chairs in mirth at my earnest, but woefully naive, attempts to 'tidy up' their communications!

So if this book leaves you wondering why the Group often present the same piece of information in different ways; why they often repeat themselves three times, or even why they construct their sentences in such a labyrinthine fashion, take it from me, these are no accidents. Indeed there is a definite purpose to their unique method of communication.

It took me a long time to understand it, but I now fully accept that it matters not whether the Group speak in a manner to which we are accustomed, or even whether we comprehend their words on a conscious "human" level or not. What matters is that they are planting "seeds" in our biology as well as our "unconscious." This is the purpose of their communications; it is their way of helping us to remember.

Given the many frustrations I encountered throughout the editing process, I cannot truthfully say that I have enjoyed editing this book. But, of course, that's only the "professional me" talking. Speaking on a purely personal (and, of course, spiritual) level, I wouldn't have missed this opportunity for the world.

I feel immensely privileged and honored to have been offered this opportunity to be challenged, frustrated, bemused, but ultimately illuminated by such a close encounter with these loving messengers from Home.

Thank you, Steve, and thank you, the Group, for blessing me with this task, and for helping me to discover this part of my contract.

Sandra Sedgbeer

A Message from Lee Carroll, author of the Kryon Material

I'm biased. Steve and Barbara Rother are friends and co-Lightworkers. But I can tell you this, the book you hold in your hand could be called OUR JOURNEY instead of RE-MEMBER. It documents not only profound information for the planet, but also the LOVE journey of two enlightened and high-vibrational people. Read it and celebrate the lives of two loving individuals who have learned to balance their duality within the new energy on planet Earth. Their message is for ALL of us.

In Love,

Lee Carroll

A Message from Doreen Virtue, Ph.D., author,
"The Lightworkers Way," "Divine Guidance," and
"Healing With the Angels."

Filled with uplifting messages of empowerment and Divine love, 'Re-member' is heart-opening and overflowing with healing wisdom and practical tools.

Doreen Virtue, Ph.D.

A message from Ronna Herman,
Author of "On Wings of Light" and "The Golden Promise",
Messages of Hope and Inspiration from Archangel Michael

I have followed Steve Rother's work for a number of years now and have observed the messages given to him from "the Group" from inspired information for human evolution into the profound truths of the Cosmos and a map leading "Home." Reading his book Re-member gives me the same all-encompassing feelings of overwhelming love and peace as when Archangel Michael over-Lights me and radiates his loving messages through me.

I am gratified that, even though the style and information given to Steve are somewhat different than what I receive, the messages never contradict each other and are always complimentary and validate the Universal truths we are receiving.

Thank you Steve, for giving us this treasure house of wisdom.

Ronna Herman

The Grand Game of Hide and Seek

Let us take you on a journey . . .

Let the Grand Game Begin. . .   Brenda Pakkala

All of us here are gathered in a meadow at the base of a mountain. We are Home and we are all playing together in perfect love. A dear brother approaches and says, "Would anyone here like to play a new Game?"

"What kind of Game?" You ask. "Is it like the ones we play now?"

"No," he replies. "It is like nothing we have ever done before. It will be an elaborate Game with many props and disguises. We will wear veils so that we can no longer see or re-member our true nature, even the veil will be hidden from view. Then we will start the Game and begin to re-member. This veil will be so effective that we will forget not only who we are; we will even forget Home. Even as we pass each other on our paths, we will look into each other's eyes and not recognize one another. The veil will be so effective that many will look around at the props and disguises and truly believe that is all there is. We will retain all of our powers, yet we will not re-member how to use them or that they are even there.

"The Game will be played in phases and before we start each phase of the Game we may place as many reminders in our path as we wish, to help us re-member. Be advised to place many reminders, for most of us will rationalize them away easily. We will choose the time and place of our entrances and exits on the Gameboard. We will also set up circumstances and lessons we wish to complete while we are under the veil. A tally will be kept and points will accumulate from one phase to the next. This point system will only be used to determine what will be included in our next phase. We will not be able to remember from one phase to the next, yet, once mastered, certain attributes may be carried forward into the next phase. We will always carry your core essence and personality, yet we will not re-member that it passes with us through the veil into every phase.

"Humor will always be a reminder that passes unchecked through the veil, and if our guides find us getting too serious they will tickle our funny bones as a reminder that this is only a Game. There will also be many masters available along the way to help us if we wander too far from the path. Oh yes, I almost forgot an important part of the Game . . . at all times there will be Free Choice. We will have complete choice in all matters, we may even choose not to play the Game, or to call in a substitute. We may choose to hide, or we may choose to seek, it is entirely up to us.

"On the Gameboard there will be polarity. This has to do with the mechanics of the Gameboard itself, and will be a necessary component as it provides the contrast needed. However, polarity will taint our vision. Through eyes tainted with polarity, we will perceive things as Up or Down, Light or Dark, Good or Bad, Love or Fear and Right or Wrong. Do not let this fool you, it is only illusion.

"We will all leave our higher aspects of ourselves in a special place for the duration of the Game, otherwise the Game would be much too easy. Our higher selves will be available to you at all times. Our challenge will be to learn to access it, and to re-member that it exists as part of ourselves. We may choose special loved ones to ride on our shoulder to advise us during the Game. Again, re-membering that they even exist will be a big part of the Game itself.

"The goal of the Game will be to see how many of us can re-member who we are, where we are from, and what powers of creation we have. Once we re-member, then you may re-merge with our higher selves and re-create Home on the other side of the veil to demonstrate that we have fully re-membered.

"So who wants to play?"

How it All Happened

*D*uring the summer of 1995 I was content playing the role of a General Building Contractor in the San Diego area. I had spent many years of my life looking for that elusive niche where I might actually enjoy what I did. I thought that money held the key. My thinking was that even if I didn't like the work, as long as I made enough money, it would be worthwhile. And so it was that my life always seemed to go in circles. Looking back, I can see that I was only a mediocre contractor because it was not my passion. It was what the Group now calls a misdirection of energy.

I had always been interested in things like afterlife experiences and astral projection. I even read some of the books by Ruth Montgomery and Edgar Cayce. For me it was all fun but I really didn't put much into it. It was only entertainment. If you had told me in 1995 that I would be traveling to the United Nations to channel entities from the other side, I would have thought you crazy. It was fun to study things outside myself, but to actually use them in my life was far from any reality I could have imagined.

Then my life began to shift as my wife and I found ourselves reaching for higher truths. I had found a book on channeled information entitled "Kryon . . . Don't think like a Human." It hit home for me and served to help me awaken to another part of myself. At first my wife, Barbara didn't resonate with the book, instead she studied Religious Science. Even though we were not on the exact same path, we were both moving forward and had begun the awakening process. For me the Kryon material was my "wake up call."

On New Year's Eve in 1995 Barbara and I decided to do something different. We attended a workshop designed toward releasing the old and setting our intent for the New Year. Then at 5:30 am (YIKES!) we would meet at the beach over a bonfire, where we would burn what we chose to release while declaring our intent for the coming year. It sounded like fun. I had no idea what I was about to set into motion.

At the bonfire, each one in turn did a ceremony around releasing. I felt out of place. It was obvious that everyone else knew all the right words and actions. Me? I was just learning to meditate, and I didn't even know if I was doing that right. But when the torch was passed to me, I felt a strange sense of calm. I became aware of laughter from somewhere over my shoulder. I remember thinking it was odd that others were on the beach at this hour and I thought it a bit rude that they were obviously laughing at me. I placed my sheets of paper in the fire to release the old and promptly raised the torch to declare my intent for the new year. Trying not to sound foolish, I opened my mouth to speak and my brain began the desperate search for a metaphysically correct phrase. Suddenly that same wonderful sense of calm settled over me again. To my surprise, tears began to flow as I heard my own voice declare to the Universe:

"I choose to be a Lightworker."

I really had no idea where those words had come from. The word Lightworker was not in my vocabulary and no one else had used it that night. Even so, there was a wonderful sense of familiarity with the word.

The others helped set my intent with their energy by yelling an enthusiastic HO! As the reverberations died down I became aware that somewhere over my shoulder there were others supporting what I had just said. I turned to look, but the beach was empty. This was my first actual encounter with the Group.

My life took a new direction that morning. I didn't know it at the time, but as the Group explained later, I had just stepped into Plan B.

Shortly after, I connected with several others who had formed a Kryon message board through America Online. At the time I was totally unfamiliar with the Internet but found it falling into place easily. This was when I first realized that I could express myself through writing. Since contractors don't get much chance to express themselves in this fashion, it was a whole new experience for me. I was like an eager child taking it all in. I wanted to know everything and wasn't afraid to ask.

After a short time of soaking up information, I got a very strong "knowing" that there was something specific that I had to do. Images began to form in my head of everyone holding hands to focus their intent. Suddenly I "knew" that by combining our energies together, even this small group could make a big difference in our world. I shared this with the others on the Internet and suggested that we do a meditation online for the healing of Mother Earth. The answer came back quickly and decisively:

"Good idea, Steve. Let us know when you have it ready."

There was the laughter again. Thus began the Internet meditations known today as the Beacons of Light and the monthly Reminders from Home offered from the Group.

From the moment I started writing these meditations I was aware that I had help. Although it was not as strong, I felt the same warm sense of calm settle over me as I had experienced on the beach. That was when I really understood that I was receiving help with these messages. I began a somewhat skeptical dialogue with these beings who seemed to be sitting somewhere just over my shoulder. To my surprise, there appeared to be several of them, yet I had a hard time telling them apart.

My very first questions to the Group were: "Who are you, where are you from, and what dimensional level are you?" I thought especially the last question would impress them somehow. I read somewhere that if the entities you are channeling won't tell you who they are, then they are the 'bad' guys. Besides that, I wanted to know if they were from the Galactic Federation, the Elohim, the

Angelic Realm or Sirius. The answer came with the same wonderful laughter I heard that morning on the beach.

"You have asked three questions - - - we will give you three answers:

It's none of your business.

It's none of your business.

And it's none of your business."

My first reaction was to go into fear thinking if they refused to identify themselves, they must be 'evil'. I informed them that I would not be able to channel them if they would not give me further information. At that point the Group told me something that I will never forget. They said very simply:

"We honor your choices."

With that there was a strange silence like I had never heard before.

Over the next two days, I felt empty; as if a part of myself had been cut off. Finally, in sheer frustration, I went back to them and asked why they had refused to identify themselves. They lovingly told me that this was my lesson in discernment. They asked that I accept or reject each message for the love content of the message itself, and not because of some name or label placed upon it. They said this was the first of the tools for living in the higher vibrations of the new planet Earth. Discernment is a way of making life choices without judgment. They gave me the following as a guide to applying discernment:

If it pulls at your heartstrings, then take it as your own. If there is anything less, then leave it without judgment, for it was simply placed there for another.

They mentioned that we humans give away our power much

too easily to impressive names and titles and that their message was to help us re-member our own power.

From that point on I began writing the messages from the Group, although it would actually be several months before they would even allow me to call them by that name. In the early writings I never talked much about the Group except to call them my Guides. Then, after a few months of writing the monthly messages, I was trying to explain the origin of this information and I referred to 'the Group' sitting just over my shoulder. They reluctantly agreed to this and finally I had a name I could call them. The name stuck.

The very first word to come through from the Group was:

"Re-member."

I can't tell you how many e-mails and letters I get telling me that there is no hyphen in the word re-member. The Group offers it this way for three reasons:

1. The object of the Grand Game is to re-member who we are and why we came here.

2. When we re-member who we are, we can re-member our powers of creation and begin creating Heaven here on Earth.

3. The process of re-membering has to do with re-integrating the various parts of ourselves. They also say that the way to set this process into motion is to re-unite with the members of our original spiritual family.

This is happening to all of us now. In fact, it is the reason you are reading this book.

The Grand Game of Hide and Seek was the first of the messages I received from these loving entities. The opening chapter you just read gave a view of life on Earth from their perspective.

Over the years I have been bringing in these messages, I have learned to look through their eyes and to see things from a perspective of pure energy. When we get used to looking at ourselves in this manner it becomes much easier to see our true path.

This book began as a series of monthly Internet meditations called the Beacons of Light re-minders from Home. These divinely-inspired messages began in February of 1996, and continue each month to this day.   You can read past messages online or join us in this celebration at

http://www.Lightworker.com

At this point in my life I went on reaching for higher truths and was very fortunate to be able to travel with Lee Carroll and Jan Tober for the next two years, as they presented the Kryon seminars.   It was this venue, and the supportive people that traveled as part of this team, that first began to show me that I was a facilitator.   Lee Carroll has been a mentor for me as I learned to trust myself in bringing in this information.   Today I am very honored to also call him a dear friend.

The Group says they are here because we have asked to move forward.   They offer us information for our own discernment and empowerment so that we may be able to adjust to the higher vibrations of planet Earth as we move into what they call "Plan B".

They are not here to teach.   That was the old way, they say. They simply offer information, and ask each one of us to filter it through our own discernment, and take from it what is ours.   In their words:

We will not tell you anything that you do not already know. We are only here with big hugs and gentle nudges to help you re-member your true power.

The information in this book is placed here for your individual discernment.   It is no accident that this book has found its way to you. Take what resonates within you and allow yourself to re-member as they take you on this joyful journey. The Group often reminds us that it is only information and is useless unless we take our own power and incorporate it into our daily lives. The Group will not answer your life questions in this book. Rather, they will help you to find your own answers. The Group says that we have passed the test and are now creating Heaven here on Earth. By blending our true power and taking responsibility for our own reality we will do just that.

Welcome Home.

Chapter I

Help!

What's Happening to Me?

Help!

*T*he new call of the Lightworker seems to be: "WHAT'S HAPPENING TO ME?"

This is a time in our history that we have never before experienced. As the vibrations of the planet increase we are experiencing new phenomena everyday and attempting to incorporate these experiences into our lives. Sometimes it actually fits and makes things easier. A lot of the time these experiences seem to be the cause for yet more questions. Like most of us, you may have attempted to go back to your old ways, only to find those doors forever closed. I can also tell you from personal experience that beating your head on the door does not make it open. For me, it has been an interesting experience because I view things differently than most. I ask for doors to open and all I can see are doors closing. The Group laughs a lot about this because they say it is the only way they can get my attention sometimes. We are now working on better communications in that area. The Group tells me it's all a matter of perception. They laugh, because instead of walking into heaven, they say I have a tendency to back out of hell. I love it when they laugh.

The Group:

It is a true honor to address this gathering. The Lightwork you are doing has implications that extend far beyond your scope of understanding. We are honored to be a part of this work. We are here to help you re-member your magnificence. As you step into your true path we experience tremendous joy. There are connections that extend to all things before you and what is perceived as separate is actually one and inseparable. In that aspect we are very much a part of you. As we embrace you, we feel the love. We like that feeling.

THROUGH GLASSES TAINTED WITH POLARITY

Due to the nature of the Gameboard there is in place a mechanical grid that presents itself to you as polarity. This attribute enables you to interact with your surroundings in a way that would otherwise not be possible. It is this same polarity that also causes you to see things as separate from yourselves. Because of polarity, what you perceive as opposite ends are actually two small segments of a larger circle. What you perceive as light or dark are only varying shades of grey. Similarly, as you remove judgment from your vision there will no longer be a need for right or wrong. It is important for you to grasp this basic premise at this point in your evolution. Your sight is beginning to develop and soon you will see well beyond the scope you now call reality. As this progresses it is important that you become conscious of your point of perception. If you wish to change the world, you must alter the point from which you perceive it. These are the subtle, yet powerful, changes within.

THE DARK ROOM

We have likened the Game you now play to being in a dark room. In this room are gathered many others. Each one is searching individually for the passage Home. The root motivation for all actions in your Game can be traced to this one desire to find your way Home again. In the dark room everyone is looking for the light. The challenge presents itself because you are unable to see one another in the absence of light. This is the mechanics of what you call the veil. In your honest efforts to find the door you often bump into one another. These bumps are often perceived as assaults and taken as setbacks. Such is not the case as this is an effect of polarity on the Gameboard. For this reason, you will find forgiveness and unconditional love the foundation of all healing. If you forget who you are at any time, please take a moment and look at yourself through our eyes.

In this dark room you occasionally glimpse the light that tells you Home is near ... and you re-member. As you move in the direction of that light it mysteriously eludes you, leaving you once again in the darkness. Such is the nature of the Game. We now tell you that what you perceive as the light of Home is, in reality, a reflection of your own inner Light. This is the reason it disappears from view as you approach. The door leading to Home is within. This is the basis of what you refer to as Lightwork. Find the Light within and make space for it to shine in all areas. Do this and all else will come to you. Those who refuse to go within, go without. It is the attribute of polarity that makes it difficult to see your own Light. In your current vibration it is not possible to see yourself as you truly are. For this, you have developed the tool called a mirror. Even as you use this tool you become aware that your reflection is reversed and it is not a truly accurate representation.

LEARNING TO SEE THROUGH NEW EYES

The tools that you placed for yourself on the Gameboard for this purpose are the eyes of each other. When you make the connections that allow you to see yourself through the eyes of another, then you have an accurate representation of God within. Often this is not what you expect to see and so you resort back to the familiar mirror, even though this may be a painful experience. For those with the courage to view their reflection fearlessly and without judgment, the rewards are great. Once you see with true vision, you understand that you are neither right nor wrong ... you simply are. Those accepting the challenge of this work will move quickly to the next level. In the process they will experience many rapid shifts. What once took years to accomplish now takes only hours. This is accurately reflected in the technological advances of your world.

It is these rapid shifts in perception that often cause what you see as areas of distress. You have spent much time viewing your world from a singular point of perception. Suddenly, through your intent to move forward, your point of perception has changed. Every morning you awaken, sit on the edge of your bed and inventory your surroundings. You get used to the way the morning light reflects off of each item. You tell yourself everything is as it should be. You label yourself "happy." Your world is in order. Then one day you visit a friend and find that their bedroom is much to your liking. You return to your own space and move the placement of your bed. The next morning, upon arising, you feel shaken and unsure as you now view each item in your existence from a new perspective. Your world has changed, and it is this change that is at the root of your fear. It has begun in earnest.

ADJUSTING TO THE NEW VIBRATIONS

Such is the case of the Lightworker. You have expressed intent to the universe to move forward into enlightenment. In the previous illustration, you not only moved your bed to a new position, you actually moved it to a higher dimension. From here everything previously familiar now has the strange appearance of props on a movie set. You may find yourself examining things as if for the first time. Lightworkers are experiencing what they perceive as difficulties in several areas of their lives at this time.

Relationships must go through periods of adjustment as each one learns to see through new eyes. Long-term relationships will particularly appear to be shaken as each one reassesses their needs from their new perspective. The greatest gift you can give to yourself here is time to adjust. Short-term relationships may find it easier to adapt because they are more comfortable with constant examination of each other, and accept these changes more readily. As you begin to see yourself as a self-contained empowered being, many old paradigms begin quickly to fall away. This may have the illusion of creating problems in your life.

This scenario will play itself out in many other areas of your life as well. Work that had been successful in the past may no longer be valid. Areas of interest will fall away as new ones emerge. Friendships built on old beliefs may not survive. You may find yourself being called to a new place to live. Lightworkers resisting the transition will not be able to reverse the effects. It is not possible to go back. Some will see themselves as standing still and place judgment on that, thereby blocking their progress even further. Lightworkers may perceive themselves as experiencing difficulty on many levels. Fear not, it is only your judgment of yourself that makes it appear so to you in this fashion. In truth, these shifts will move you forward into a higher reality and closer to home. Release the judgments and you will see the truth. These shifts are steadily moving you into your joy. What you have done is ask for the door to open. This has been granted to you. This request set into motion the accelerated process. This process is the gift. The road to enlightenment has no speed limit. The Earth is moving forward and raising its vibration at an accelerating rate. It has begun in earnest.

There are many stories in your own history that illustrate this point. Among these are the tales of the Shiva and the Shakti. The way to progress is to ask the Shiva to come in and destroy all of your surroundings. Only by releasing them can the Shakti come in and create anew. The bright leaves of Spring must be preceded by the withering leaves of Fall. In every season there is a purpose. For the Lightworker this is the season of Fall. Old paradigms must fall away before new ones can sprout. This is the time of the Shiva. It is only your perception and your resistance that make certain aspects appear as problems. Embrace this time and give thanks, for it is literally an answer to your prayers. As your perception continues to shift you will look on this time with great joy.

As this shift progresses, time will be vastly misunderstood. All your relationships with time, as you knew it, will need to be rethought. This can be viewed as an area of hardship for you. This need not be the case. It is important to understand that you are making all these adjustments to your world to compensate for the changes in vibration. It is also important to know that those around you are still seeing things from your previous point of view.

You may find yourself waking everyday with a sense of urgency and a drive that you have not had prior. If you have insufficient outlets for **ex-pressing** the light within, this will be especially frustrating for you. It may seem that no matter how much effort you expend you still are not proceeding. At this juncture, applying more energy to move forward will only be met with more resistance. You have placed blocks in your own path that will remain until you understand their meaning. We tell you that there is ample time for all. To force the issue of time will only serve to block you in your overall progress. Give yourself the gift of time and judge yourself not.

NEW TOOLS FOR HIGHER VIBRATIONAL LIVING

There are tools for this shift that are yet to be discovered. We will speak of some of these at this time. As all of the work you have agreed to do has its basis within, look there for the tools of transformation. Many of you have worked to rid yourselves of emotional scars. This is highly honored as it allows you to comfortably carry more Light within. You have reached a level of comfort where these scars are seemingly healed, never to bother you again. We see this as being much like a tube through which you carry this energy through your biology and into the Earth. Emotional scars have the effect of restricting the amount of energy that can flow through this tube, much like stepping on a garden hose. You have become accustomed to carrying a specific amount of energy through this tube. Most of you have done just enough emotional work to allow for this flow comfortably, yet even as these wounds heal they leave behind scar tissue as a reminder. Now you have expressed intent to take the next step of becoming Lightworkers and playing an active part in the next stage of your evolution. This is a step from your "Plan A" script to a conscious decision to walk into "Plan B". From our perspective, this is much like placing a large funnel on your head in order to bring more Light through this tube into your biology. Because the amount of energy has increased, the scars that you thought were fully healed may now once again become agitated and painful. This has the effect of pushing 400 volts of electricity through a wire that was designed to carry 10. Every bend and nick in that wire will begin to heat up. This pain appears to your limited vision as a setback. It is, in fact, a gift, because it allows you to work through the remaining scar tissue at a greatly accelerated rate.

There are many of you who will experience grounding problems as you begin to carry this heightened energy. Your spirit will want to leave your body at every surge of this energy. The work you have chosen to do is only possible by maintaining adequate grounding. We have spoken before of the need for grounding. Listen intently as your biology speaks to you. It will tell you what you need for grounding. This is a time when your diet may change. We suggest that you make space for that to happen by removing judgments.

Connection to water in all forms will be essential. Water is a special form of energy. The electrical and magnetic properties of water have yet to be fully understood. Know that this can be very helpful in your efforts to achieve grounding. Bathing in salt water will be helpful for most. Drinking increased amounts of water will help the biology ground. With everything that enters your biology, it is suggested that you precede it with ceremony. Such is the basis of what you call a blessing. These ceremonies are helpful in engaging the biology, and are most effective when you co-create your own. Intent is the most powerful tool you have. Incorporate spoken intent into your ceremony and you then have powerful, personalized tools.

Breathing is a tool that can be utilized at any moment. Each one of you has three or four sacred numbers that you can easily discern for yourselves. Ask in a silent moment and they will be given to you. Most of you already know them in your heart. These single digit numbers are a signal to your biology, and act very similar to sacred geometry. Incorporate these numbers into a breath pattern of deep and shallow breaths, and you will have powerful grounding tools designed specifically for you. This is a realization of your own powers within. If you incorporate ceremony into this exercise it then allows your biology to be part of the process. We think it very humorous that you only give power to something if you surround it with mystique. If that is helpful, then do so with blessing. We wish to point out that it is not the ceremony that contains the power. Rather, it is the ceremony that is helpful in aiding you to connect with your own power.

Emotional grounding is a largely misunderstood area. As you move into a higher vibration, you begin noticing shifts in time. Your perception of time now takes on new meaning. With inadequate emotional grounding, your perception will be that there is no time to accomplish all that lies before you. Understand first that this is illusion. It is simply a by-product of polarity on the Gameboard. Emotional grounding will be discussed at length in future sessions. For now, it is effective to visualize a cord running from your root chakra. Much like the silver cord of life that extends from your biology, this cord has great elastic and flexible properties. It is used to ground your emotional body. As you feel yourself disconnected and fluttering emotionally, go within and check the position of this cord. Your observations will find it fluttering around, much like an uncontrolled garden hose as water spurts out in random directions. The tool to use here is to stop and breathe. Ask the Earth for permission to connect, then consciously take this cord and insert it into the Earth. Do this, and feel the sense of peace that overtakes your being. This connection with the Mother is a gift that all can experience.

Even as you travel, this cord can be used to ground your emotional body in any circumstance. Simply re-membering that it is there will start the action.

You now find yourself back in the dark room. Others bump into you as they search for the door. Some run through the crowd as their fear overtakes them. In your humanness, all you are looking for is an opportunity to feel good. When you feel good it is a re-minder of Home. Everyone deserves the right to feel good and to re-member Home. These are the crystals that you have placed in your own path to mark your way. This is the motivation behind all action of those around you. Judge it not. Understand that judgment is an effect of polarity on the Gameboard and is only illusion. Do this, and you will find the key to the door Home. There is but one way to find the door in the dark room and keep from hurting each other: Hold each other's hands.

Bless each step on your path as it unfolds, no matter from which point you perceive it. This is a very special time and you are honored beyond your understanding for every step that you take. You have asked to move forward and now……. it has begun in earnest.

Dark Room Exercise in the Scepter of Self Love, Milwaukee, Wisconsin

It is with great love for you that we ask you to treat each other with respect, nurture one another and play well together... the Group

Question: How many Lightworkers does it take to find the door in the dark?

Answer: All of us.

This time presents many challenges for Lightworkers. For many it is not an easy path we have chosen. I often want to scream HELP I'M MUTATING AGAIN! When I am in the middle of the transformation it is not so easy to see the larger picture. It's not always easy for me to re-member that I am in a dark room. This is the reason we have each other. This is why they suggest we hold hands. Together we have exponential powers of creation. Together we can change the world . . . one heart at a time . . . starting with our own.

Chapter 2

"Plan B"

Re-membering your spiritual family

Barbara and Steve in Salzburg, Austria

Plan B

The Group often speaks about the changes we humans are experiencing as part of the shift. They say we are in the midst of
implementing "Plan B". They have said that the object of this elaborate Game was to see how many could re-member who they were,
and their powers of creation. When this was accomplished we were
to re-merge with our higher selves and use our powers to create
Home on this side of the veil. For the longest time it did not look as
though there would be many who would accomplish this. Only a
few masters would fully re-member their powers. Although those
were great accomplishments the masses never got the overall picture.

The Group describes us as being finite pieces of the infinite
Creator. Infinite simply means not having a beginning or an end.
Being infinite, the Creator can easily accomplish everything with the
one exception of observing itself. The idea of the Gameboard of
Free Choice was to send off sparks of the Creator that had a beginning and an end to play the Game. The sparks were made in the
exact image of the Creator and given Free Choice in all matters.
Therefore, it was the perfect place to observe the image of the
Creator as reflected through the choices made on the Gameboard.
Since the sparks were actually parts of the Creator, they still had all
the powers of the Creator. They had only to re-member how to use
them and that they were even there. The big difference was that the
sparks were finite and after a bright display they soon faded off to
reunite with the circle again. The sparks are us.

Since we are finite, and therefore have both a beginning and an
end, the Gameboard we created is also finite. Although the Group
has not mentioned it, I imagine the beginning of the Gameboard to
be the Big Bang. The end was designed to occur right about now.
This was to be Armageddon - the much-prophesied end times, as

predicted by Nostradamus, Edgar Cayce, the Hopi Indians, Revelations in the Bible, and more recently, people like Ruth Montgomery, Seth, and Gordon Michael Scallion. When that information was originally channeled it was very accurate. It was pointing the direction we were headed at that moment. In fact, it was this information that allowed us to alter the outcome.

As things were positioning for the Game to end, it was noticed that many of the players were beginning to awaken and re-member. Although it was very last minute, this brought about great excitement universally. It meant that the Gameboard of Free Choice just might make the evolutionary shift and forever change the paradigm of life throughout the universe. This event brought out a multitude of beings from all over the universe to watch as each act of this play began to unfold.

Several of our parental races began showing up to help us in this time of transition. The Group has pointed to the humor that was felt over the fact that some of these races thought us insignificant until the shift began. Now, they say there were even power struggles to see who would get to help us the most. The Group has stated that there are so many onlookers here to view this event of cosmic proportions, that what we call parking is really beginning to be a problem.

Before we came here we designed elaborate plots to facilitate our lessons on the Gameboard. We had the system of Karma in place to keep score so that we could easily decide what we were going to work on the next time around. It is because of Karma that we have often incarnated in groups or families. If you and I were to have a chance to work off Karma, it made sense that we should be on the planet at the same time and in roughly the same location. Over time, it became possible to complete Karma with your original family and then join with another. This mingling process continued until today, which is why very few of us re-member anything of our original spiritual family.

When scripting our parts we asked this person to play our father and this one to play our Mother, as they were perfect for the role. Conversations went like this: "My very dear brother, could I ask you to please be my business partner when I turn thirty one? Can I also ask you to help me repay a debt by stealing all the money? Would you love me enough to do that?" Unlike previous phases of the Game, we knew this one would be different, as it might very well be the final act. Because of this we set up many alternate plans. After all, with Free Choice overshadowing everything, we never really knew what direction things could go. Because of Free Choice, all contracts are only potential contracts until they are actually taken. It was therefore necessary that the important roles had many back-up plans to ensure that they were played out in all events.

We knew that even though it didn't look like humanity was going to graduate we must still make provisions, just in case. Suddenly we find that our wildest dreams have come true and the most wonderful events are now about to unfold. Instead of the Game ending, humanity is now at the brink of the next evolutionary step. We, as players on the Gameboard, suddenly found ourselves in an awkward situation. Even though we had rehearsed our parts and knew our lines well, we suddenly found ourselves holding scripts that were no longer valid. In an instant the Game changed directions. Without a single word being uttered, all of us dusted off our scripts and flipped the pages frantically until we saw the heading "PLAN B".

The Group:

We welcome the opportunity to bring information to this beloved gathering. You are honored more than you know. Your willingness to play the Game behind the veils is empowering the universe in ways you are unable to fathom. Your perception is limited because of the veil you wear. You see us as powerful advanced beings, far beyond your stage of evolution. We tell you once again that it is you that are the honored ones. For you have come to this grand Gameboard as masters and agreed to hide the truth from yourselves for the good of all. We are not here to teach, and we do not lead you into a new level of existence. These things you have done for yourselves. We are here at this time because you have asked for assistance in this transition, and this we give you with our thanks. You are beginning to hear more clearly the calling of your own heart and we are here to validate this. Many of you have intended to move into the next level of existence. We cannot fully describe to you the greatness of this act. We are here at this time to answer your call and to help you to re-member your own heritage and powers of co-creation. These are the tools that will readily move you into this next level. While moving to this next level, it will be helpful to be aware of the changes that lie before you. At this time we will share some of these with you.

THE TRANSITION TEAMS

The planet is rapidly approaching critical mass. This coincides directly with the number of souls incarnate on the planet at any one time. When this critical mass is reached there will be many leaving the planet. This is in all appropriateness and is for the highest good of all. In preparation for this it will be of great use to address techniques of transition. On the planet there are teams that will form for this purpose. These are very special healers who carry information with them for those who will be transitioning and their families. Many souls leaving the planet are often unprepared for the wonders that await them on returning Home. Education on a mass scale and individual transition teams may make this thing you term 'death' much more of the joyous homecoming than you presently perceive it to be. When beings transition without an understanding of what to expect, they often stay in levels much longer because of the fear they hold of the unknown. This information is now available and will be offered more as events unfold. You all would be well advised to provide space for these master healers to take their place and lend them your support. They are here to play a key role in the development of your next level of vibration.

As we have mentioned prior, this is a time of change for you on many levels. Your outer ethereal bodies have been in progress for several years, and now the changes to your biology have begun. For the past **fifty-three** years you have been accumulating seeds in your biology that are now beginning to sprout. These seeds you have accumulated through vibration in preparation for this movement. They have been presented through the means of sight, sound, smell, and absorption. What we now refer to as seeds, others have termed activation codes. This is the reason for the phenomena you call crop circles. It is also the motivation behind the increased interest in what you call sacred geometry. Number combinations are often used by spirit to activate these seeds. This is the reason many are now seeing repeating numbers appearing in their fields. These are the master numbers that trigger your biology to activate seeds that have been lying dormant for a very long time. When this occurs, fear not, as it is a signal that you are receiving the gift. Bless the gift and know that each time it happens spirit is standing directly next to you. From our perspective these might be called "Angel Blessings". They are preparing you to re-unite and once again walk in the vibrations of Home while still on the planet. When these numbers enter your field we caution you to simply accept them, rather than sidetrack the energy by attempting to attach a meaning to them. These seeds are now activating and are beginning the changes to your biology that will enable you to continue on the Gameboard of Free Choice. It is a gift you have earned. Accept it with joy and blessing.

FINDING BALANCE IN DIFFICULT TIMES

These shifts within your physical bodies are of concern to many, *therefore we will speak to you of the tools available. Your choice to make this shift has caused you to carry increased amounts of energy through your biology. If you are not able to ground this energy it may cause discomfort in both your physical and emotional bodies. We have given grounding techniques that can be most helpful. Grounding to the Mother will be very helpful, not only to you but also to the Earth. You are both in the process of raising your vibration, and connecting you helps to balance each other. The shallow breathing that is so common in your world often perpetuates many of these grounding problems. Breathing consciously to a pattern of deep and shallow breaths, intuited through your own connection to spirit, will balance the biology and ground the energy. Increased amounts of water absorbed both inside and outside the body will also be beneficial. When possible, bathe in water with soluble salts.*

Your diet may need adjusting during this transition. Intending with ceremony that food be used only for your physical body will diminish the need for emotional eating. This is a time to be very sensitive to what your body tells you it needs. Practice the art of listening. Trust what you hear and release your judgments long enough for your body to balance. This is not a time to force your biology into a shape you deem to be pleasing. Rather, we suggest that you allow your biology to balance itself in the most efficient way possible. Your body is the perfect housing of your spirit. It is helpful to keep your muscle tone at comfortable levels, however, since undue stress during the vibrational shifting would not be in your best interest. Here the key is balance and listening to the body.

The use of vibration will come to the forefront of your medical sciences in the near future. Seek out vibrational healers, for there is much information already available on this subject. Placing colors within your field that balance your mood is a tool that you can use to control the flow of energy through your body. This may require a rethinking of the use of color and will require more individual expression concerning the use of colors. Vibration received through the sense of smell is also a powerful tool, thus many are now becoming extremely sensitive to aroma. This is a signal to be discerning about what you allow into your senses. Become aware that all vibrations entering your senses have a direct effect, not only on your biology, but also on your emotions and spirit. You have another sense that you are not aware of that will become important. This is something that we will call only absorption. It has to do with the manner in which you absorb energy. As you become more aware of how this works you will find many of the keys that you seek. The sights, sounds and smells of the Gameboard have rarely been regarded as tools, yet we tell you now that these are great tools for what you term ascension. The art of vibrational healing will move to the forefront of your sciences in the near future.

Your patterns of sleep will continue to change as you move forward. There may be changes occurring in both directions and neither are causes for alarm. A gradual move to shorter naps may help with this shift. Simply allow your body to find its own balance in all areas. There are some of you measuring your progress to ascension by the physical changes you are expecting. When these changes move opposite to the anticipated direction it causes you undue stress. Release the judgment and know that all is happening for your highest good.

Each one of you is in the process of raising your vibratory level. As this proceeds you are becoming aware of the many other levels of existence. What you would term "alternate dimensions" are beginning to present themselves to you. You will notice them first as brief shadows passing through your field of perception. As your sensitivity becomes more acutely tuned, you will be able to focus on scenes and beings that are seemingly occupying the same space you occupy. In most cases these will first appear with an obvious absence of color. There is much to be gleaned from these other *levels of existence and more information will be forthcoming soon. In fact, many of you have been studying this without the direct knowledge that this is what you were studying. This knowledge will answer many of your questions. We bring you this information now so that it is not accompanied by fear when it presents itself. Fear is the one emotion that can easily retard your progress in many areas.*

RE-MEMBERING YOUR ORIGINAL SPIRITUAL FAMILY

The changes that await you as you move into the creation of Home on Earth are numerous. Many are beyond your comprehension at this stage of your development. We will address these another time. There is one event that has begun on a global basis that is responsible for much of the vibrational advancement seen on the planet today. It is with great reverence that we now speak of this. There is an awakening process now in motion that is rapidly changing the face of the planet.

These changes are in preparation for the next step of your evolution. This is a plan that you have written for yourselves that is now firmly in place. As you know, we are here to help you re-member. We have told you this is not only to re-member who you are, but also to help you reconnect with members of your spiritual families. This has started and you have now begun to connect in much larger numbers than we first thought possible. This is key to your movement at this stage of your development. The scenarios you set for yourselves prior to entering the Gameboard of Free Choice had an alternate plan. The grand plan stated that if the awakening began, you would locate the members of your original spiritual family and re-connect. This would allow you to balance your own vibrations in a re-membrance of Home. By looking through the eyes of those within your original spiritual family you could most easily re-member your true heritage. Even brief interludes with these family members will cause the internal codes to take hold and move into action. *This is happening on a grand scale now. On the Gameboard, the original plan was to pass each other on your paths, never recognizing one another through the veil. Now, as you pass one another, you still do not recognize the physical, yet you instantly know the energy. It is a reflection of Home. When you view yourself through their eyes everything is in perspective. Please do not cling to this process thinking that you need these people to be whole. You are complete within yourself and these connections are to activate your advancement. These connections are the greatest gift you have placed in your own path. Fear them not.*

The re-membering of spiritual family is the most effective activation that is available to you. It is your greatest tool. The most efficient way for one to re-member who they are is to re-member their origin. These families are much larger than you might first imagine. There are many families on Earth at this moment, yet all of these are directly traceable to the original seven houses. It is not ours to speak of history except where it directs your step at this moment. There are others in place for that purpose. We speak this only so that it has a chance to resonate within your heart. We tell you that when a family member enters your field you know their vibration well. They carry the feeling of Home and you will find yourself drawn to them. Most often they look very familiar to you, and though you may not know them, you feel you know them well. It is humorous to us, because indeed you do! Feel the joy when these people are in your field. These are crystals you have left for yourself to mark your path on the Gameboard. This is why they pull so deeply at your heart-strings. Seek these people out and make space for them to interact with you at every opportunity. They will lead you Home again.

The Earth is also a part of this family vibration. It is your connection to the Mother that will balance you at all times. This was your original vibration and re-membering it will help both of you move to the next level. Work together with the Mother in all areas, and send her healing energy as you would an ailing parent. This energy will return to you as quickly as would a reflection, because of the deep connection you share.

MICHAEL - THE FAMILY CONNECTION

We will tell you that the act of re-membering original spiritual family is not only happening on your side of the veil. We, too, are also in the process of re-uniting our energies. There are many teachers on the planet at this time and they can be easily grouped into families of vibration. This is a difficult concept for us to present because of your predisposed concepts about this side of the veil. The Keeper views us as "the Group" and has even called us by that title. We tell you now that all on this side of the veil are a group of one form or another. There are many factions of these groups, yet all carry the vibration of their original family. We now will verify something that you already know in your hearts. We are of the original family that you would term Michael. Your perception of the angelic realm does not always allow us to come to you in our true form. We make exceptions where necessary. We tell you that the angelic realm is real, and you will be seeing much angelic representation in the very near future. This is the re-membering of family in its deepest aspect. This is who you really are.

The Grand Game of Hide and Seek on the Gameboard of Free Choice was about balance. It was designed to splinter the original vibration into harmonics. The seven, plus the original, represented eight. From there, they further split into individual harmonics on a grand scale, and then split many times again. Now, as each of these individual harmonics chooses to raise its vibration, they are naturally drawn to re-connect to their original tones. Once re-united as families of vibration it becomes easier for all to advance together as one. This is now happening at an increasing rate. As these families connect they will begin to attract other families and the next level of vibrational re-membering begins. This has the effect of a multitude of harmonics spreading out to form the sound of the original om. This is what you are experiencing in your lives as you lead the way for those raising their vibration. It is important to re-member that there are no steps on a ladder and no one is higher than another. It is simply harmonics of the original tone and all are honored for their vibration.

We are deeply honored for the work we are allowed to do in this re-membering process. It is our greatest dream that we will re-unite with all of our grand family once again. You on the Gameboard are making this possible through what you see as your daily struggle. In our eyes this is the most highly honored work in the universe. You are re-membering yourselves and us with the part of God that we are. You are loved beyond your understanding by all of your family.

It is with the deepest of honor that we ask you to treat each other with respect, nurture one another and play well together... the Group.

This information came in at such a fast pace that I found it a real challenge to type as fast as my fingers could move. When this happens they are usually showing me much more than I am able to put into words. In this channeling I got to see many visions of the future, although I was only able to put a few of these down. I will tell you this, though: it's going to be a wonderful time. Much like a great party. It is a re-union party that will end all parties. Seems like the entire universe has been invited for this one!

I say again that I am truly honored to present this information. It is my greatest pleasure to carry this information in the forms of these monthly meditations, the seminars and the private sessions. I am aware that if no one expressed an interest in these messages I would not be in my contract, so I really owe a great debt to all of you reading this for making it possible. Over the years I have always told Barbara that I wanted to have a job where I couldn't wait to get up in the morning. Well, thanks to all of you, I now have that. You see, the way I look at it, I am beginning to create my version of Heaven on Earth. I am very grateful for the part you allow me to play in this grand Game.

Chapter 3

Awakening the Master Healers

Finding your "Plan B" contract

Awakening the Master Healers

I was on an airplane to Sudbury, Ontario for a gathering of Lightworkers when I began receiving this information. As I boarded the plane, I caught the eye of an elderly man behind me in line. He was alone and had a twinkle in his eye that drew me to his stare in an instant. I knew nothing about this gentleman, yet somehow I did. This guy looked really familiar. Knowing what the Group had said about seeing familiar faces and reconnecting to spiritual family, I made a mental note and uttered a quick thank you to spirit for this gift. I boarded the plane and found my seat only to find the man passing me in the aisle on his way to the rear of the plane. As he passed our eyes met momentarily. That connection was very powerful, and though all I could muster was an awkward "hello" there was really no need for words. He graced me with his smile, and that was enough. We knew each other and both of us knew it.

I settled myself for the journey and thought nothing more of it. Halfway to Chicago I pulled out my laptop and began writing. The Group and I love to write on planes. I don't know what it is, there's just something very special about it, and I always drag along my laptop to accommodate this. Just as I started up the computer I saw a very familiar figure, as if he were standing right before me with his finger high in the air.

Some time ago I got the pleasure of describing a few members of the Group. At the introduction of "the Merlia material" presented later in this book, I spoke of an elderly gentleman with a great white beard whose job it is to simply hold up one finger. This finger is to signify when it is appropriate to release information to the collective consciousness. In many ways he is like the grand farmer. The seeds must be planted at exactly the perfect time for the germination process to be effective. It's his job to determine this timing. For a time this old gentleman had left the Group and I missed his energy dearly. They had told me before that the Group was always shifting as some members came and went. I finally put two and two together and realized that the gentleman in the rear of the plane reminded me of this grand wizard of timing.

I wrote for about an hour and as the following information came through this man kept coming into my thoughts. The airplane seats were positioned such that it was impossible to get a look back at the rows behind me, but I could sense that he was there staring at me and thinking of me. Finally, with the aid of coffee and a diet Coke, I decided to use the restroom at the rear of the plane and get a better look at this guy. I even thought that if the seat next to him were vacant I would move there. I went back as slowly as I could, checking out every seat, and was very discouraged to not find him where I thought he was sitting. I reached the restroom and luckily found it occupied. This gave me the opportunity to go to the front of the plane and continue my search. (As if I needed an excuse.) Still no luck, and as I returned to my seat and my writing this man was somehow always there on the edge of my every thought. It was as if he were watching every word as it appeared on my screen, nodding in approval.

The Group:

We are truly honored to be addressing this gathering of those who call themselves Lightworkers. This is an honored title you have claimed for yourselves and clearly identifies your origins. By using this word you have set your intent to the universe to be one of the special few. We honor your discernment highest of all because in this process lies the act of fully expressing the spirit within each one of you. We are here to help you find your own empowerment and your own truth. We offer you this dialogue for your individual discernment as it relates to your quest for higher vibration on the Gameboard of Free Choice.

There is much activity now on the Gameboard as the players are re-uniting as spiritual families. Many doors are now opening as a result of the changes set into motion by the re-membering process. The act of re-membering your true nature and your higher contract is most highly honored for it allows you to fully claim your power. It is this process of re-locating your original path that so many of you are now experiencing. The Grand Wizard of Time, about whom the Keeper has spoken, is now indicating that it is appropriate to release more specific information of this awakening in progress. With this mass re-awakening, there are many who are moving into their contingent contracts. Thus, it is now time to speak of a special group known as the Master Healers.

This is the age of movement. Moving into your next stage of evolution brings about the need for many facilitators. It is time for an awakening of the Master Healers on the planet. There are many in this family vibration that are now hearing the call to take their place as Master Healers. We are of the family of vibration that you call Michael. This family is one that houses a very large number of beings that have this propensity toward healing. Let us explain. Many of you have experienced past lifetimes where you mastered the art of healing in one form or another. As stated in the beginning, once mastered, certain traits may be carried forward from one phase of the Game to the next. Hearing the call for the last stages of the Gameboard, many of you chose to be here at this time. You are what we would call healers in hiding, for many of you did not come in this time to work in the area of healing, but rather came in with other roles to play. Now that plan B has been called into action you are feeling the pull to position yourself in those primary contracts once again. This section of the original family contains many of these masters. Those of you drawn together in this vibration have roots in this family of healers. This is **our** reason for being here as **we** are experts in the area of awakening these healers. This is **our** part of the re-membering process.

There are many who do not consider themselves healers per se, so we will define more clearly the attributes of the Master Healer. A Master Healer is one who creates space for others to feel safe enough to heal themselves. This can manifest in many areas of your Gameboard. At one end of the spectrum are those in the traditional healing arts. There is much truth here to re-member. Please do not dismiss this information without thought. Within this definition are also those that work with energy. This is a truth that has been around for eons, yet your science has yet to define it. Follow the results and this will lead you to the truth. There are also many more that we term vibrational healers. These include those who heal through the use of vibration in the form of music and art. This is important as it is beginning to emerge at this time, and will explain many of your pulls in direction. There are very many of you who have also moved into the healing arts for the first time. Most of you will know who you are. Please understand that you are highly celebrated for this decision. Honoring a pull to change your life to follow your heart is a true honoring of the spirit within you.

MORE ON THE TRANSITION TEAMS

We have begun to speak of those we have called transition teams and it is time to elaborate on this further. The mechanics of the Gameboard have eluded you for eons. When you are able to see them you will be surprised that they are so simple. We will tell you of some attributes that you are not fully aware. The attitude with which one leaves this dimension will determine the trajectory of re-entry. It is not understood yet that your last vibrations are the same ones you must pick up upon re-entering. Those who die traumatically do so to complete cycles or clear Karma. Those that die at the hands of others are instantly bonded and will incarnate together from that point forward until such time as the debt is fulfilled. Please understand that this thing you call pain is an illusion to enhance the

Game and at no time is there any real danger. As you know from your practice of what you term Astral Projection, the spirit can leave the biology at times when necessary for the facilitation of such traumatic events. Your vision of this thing you call death has begun to change only recently. With your study of after life experiences you have begun to reveal universal truths. This is humorous to us as we see what you term "after life" as a "return to life" and to your true nature. This has also enabled you to open your understanding of your own spirituality, for it reveals universal truths that cannot be denied. If you wish to study life on the Gameboard we ask you to speak to one who has returned from death. You will find peacefulness naturally emanating from the deepest cellular levels.

The systems you have devised for healthcare have grossly underestimated the importance of graceful and informed transitions. The posture of transition being something to fight against is not conducive to soul advancement. We ask you to support the new paradigm and set it into motion now. Make space for these teams to come together and offer them support, as they will benefit all. There are many Master Healers on the planet feeling the pull to be part of these transition teams. They are healers working to complete closure on the Gameboard for those leaving. They will work not only with those graduating but also the families seeking understanding and participation in the transition process. These Master Healers are highly honored for their work as they offer us the most advantageous use of the Gameboard. Through their work they make possible an easy transition without accumulating the extra burdens usually associated with this process. Upon re-entry it is then possible to set about the tasks of advancement without needing to unravel unnecessary energy tangles.

There will be many leaving in the near future. We ask you to celebrate their return Home, for they will be playing a grand part in this shift that is now in motion. You would not be able to accom-

plish this shift without the balance of their transition and the work they will be doing on this side of the veil. They have agreed to leave the Gameboard to aid in many other ways that are only possible on this side of the veil. Many will carry seeds of the new energy as they return to biology. Most will be returning as higher vibrational humans with attributes that will carry you into your next stage of human evolution. There is more to what you have termed Indigo Children than is presently understood. More will be forthcoming about this next phase of the vibration and the biological advancements taking place to facilitate this movement. Those leaving are to be acknowledged, for they offer assurance that you will meet the challenges ahead in the higher vibration. Those who leave after asking to be part of the ascension process are those honoring contracts for the betterment of the whole. They carry with them the intent to be part of this process, and so it is given unto them. They will play very important roles in the ascension process from this side of the veil where much will be accomplished.

THE MASTER HEALERS

Many of you resist the idea of being a Master Healer, for the pictures that hang on your wall and define your world declare that you are not worthy of such greatness. We ask you to keep an open mind and let these re-membrances unfold for you naturally. You may find yourself being drawn to put your hands on another with the intent of healing, only to find that you have unusual powers in this area that you have not yet used in this lifetime. You may even be aware that you have always had these gifts, yet they were always somehow in the background. Now you are feeling the tug to find out more and move into these fields once again. Honor these gentle nudges for they are pulling you to find your true path once again. It is these gentle nudges that will lead you into your greatest joy and passion as you move into your true path once again. It is time to take your place, for there will be many facilitators needed as you take each other's hands and walk forward together.

ABORIGINAL HEALERS

It is this resistance to step into your power that is now drawing many of you together to form groups of healers. This is a process that will unfold naturally as the vibrations continue to increase. The healing centers forming as a result of these pulls will enable many to assume their contracts. There are many vortexes of healing energy globally, now being defined by these groups. They will be widely known for their results and these will be their greatest credentials. Most of these gatherings will focus around one central healer who is pulling the others together. It is the task of this central figure to carefully monitor the energy distribution within these groups. As these healing centers begin to gain popularity the main opposition will be from within the group themselves. The hazards that contain the ability to derail these groups will be issues of the unbalanced ego. The primary focus of these central, aboriginal healers will be to create space where the central core of healers can balance between standing firmly in their own power and contributing to the whole.

The other task at hand for those anchoring the group energy will be to balance this energy to allow for new modalities. These healing groups will bring the availability of many new forms of healing. They will offer classes in empowerment and healing modalities as well as emotional clearing. In many ways these will be spiritual centers offering healing on many levels. Techniques in healing modalities will be offered for the emerging healers as well as spiritual contact for the emerging soul. The formation of these centers will mark the next step in understanding our true nature, for they will alter the paradigm for all things to come. Now is the time when you will look to these centers, not only for healing of ailments, but for tools of advancement. These centers will be learning centers as well.

Be advised to not confuse these vortexes with the buildings that house them. This new paradigm will be successful because of the energy mix of the central core of healers and not because of locations or buildings. Many of these centers will float for a time until the cooperative energy of the area will support them fully. Some will start supporting speakers and offering classes only. They will be just as powerful in their purpose. When we use the word center we mean a central energy vortex to which the main core of healers focus their energy. This will be the driving force that will call these groups into action. After a time, these healing circles will draw many to their diverse offerings and their success will be their most powerful credentials.

We ask you to re-member to allow space for all information to express itself. What you term traditional healing modalities contain much that is valid information. The most effective modalities will blend the physical and the metaphysical sciences. In time the postures will relax and allow more of the truth to come through easily. As the vibrations of the planet increase this will come to fruition. These centers will also be some of the first to house the coming transition teams. This will be a grand re-awakening and re-membering of those we call Master Healers. As the Master Healers once again move into the picture the healing energy will be unleashed to the planet on all levels.

You have placed many crystals on your path to mark your way. When you run across them they vibrate of Home and you feel tremendous joy. Find this joy and follow your heart, for it will lead you to the middle of your path of least resistance. Many of the bumps that you have set into motion for yourself were designed to open you to these potentials. Take them as the gifts they are and understand that the power they have is only a reflection of your own magnificence. It is now time to walk in your own truth. This can be done through the balanced ego. It is ours to help you re-member who you are. It is yours to accept that truth as it is revealed to you and walk into that contract with quiet dignity. Find that truth in your own heart and silently carry it as you would a sword of strength. Brandish it not, for to do so indicates that the power is within the sword. The sword is but a reflection of your own power and is most powerful when used to remind you of your own truth. It is time for the Master Healers to awaken to their truth again.

Sword of Truth Ceremony
at the Lightworker Spiritual Re-union
Aarhus, Denmark

"FLAVORS" OF THE TRUTH

Source is singular and all is traceable back to the prime source. Because of the diversity of your personalities on Earth there is a need for many paths to the truth. There are none that are right or wrong. There are no "clear channels" or paths to source, as all information must pass through filters and finally biology before it can be expressed. Therefore, we tell you that all information is altered slightly. There is great diversity with each one of you and therefore each one filters the information slightly differently. This is also reflected on this side of the veil, as we are much more greatly varied in what you call personality than are you. These, too, are filters through which the information must travel. In some fashion all are correct. Your task is to use discernment and find the ones that resonate within you. It is a matter of finding the vibration that most closely matches your own. We ask you to consider these filters as you would a flavoring - it either suits your taste or it does not. It is helpful to re-member at this point that you are very powerful beings who create with your thoughts. If, in discarding one of these flavors, you choose to judge it as negative energy, you actually create this and give it far more power than it deserves. Such is the case with what you call evil. It is simply misdirected energy and is a reflection of your own need for judgments. Therefore, in choosing what is right for you we ask you to discard that which does not resonate with you without judgment, for it was simply a truth placed there for another and not for you. We ask that you do not give power to something that does not deserve your attention.

THE GAME CHANGES

Being on the Gameboard of Free Choice places you in a special situation that is shared with no other species in the universe. Yours was a Game of hide and seek. That Game as you wrote it is now ending. The next phase of the Game will be played on a different Gameboard with another set of rules in place. This was a true Game of hide and seek. You allowed yourself to forget that part of the prime Creator that is within each of you. In our eyes, there were twists and diversions to the Game that made it impossible for you to find your way home again. Now, at the eleventh hour, you have begun to re-member who you are. There is no greater homage that can be paid to God than to re-member the God within each of you, and to reflect this within your own life. It is through your hard work, the work of your hearts and your souls, that this Gameboard will now move into the next phase. It is beyond our understanding to see how you tolerate an existence with limitations such as those you carry within biology. Your feelings and inner struggles within your emotional set up are the most difficult for you to change, yet these are the ones that allow you to carry the most light to the Earth with the greatest efficiency. Often it is your emotional turmoil that brings the greatest advancement for the planet. For this you are loved beyond your understanding. The many conflicting rules that have been laid before you to follow are confusing at best. Let these teach you to follow your heart in all matters and to discern everything within your own field, including rules that lie before you. These are the tools that move you into your power. You have begun this search in an effort to change the world and recreate Heaven on your side of the veil. This is your greatest accomplishment, for in changing the world you are expressing the powers of co-creation that will call for Heaven to be created on Earth. What you did not expect to find is that the key to changing the Earth lies within each one of you. We offer you this for your own discernment in the greatest of love. Changing the world lies within each of you. Change yourself and you will change the world... one heart at a time.

Extending our hands to yours is our greatest honor. With the veils you wear to facilitate this grand Game, you have been unable to see the true work you are doing. To live an existence within the cumbersome bubble of biology you wear to facilitate this experiment places you among a small number that will be revered for eternity. Your badges of courage will be with you forever more and will be honored by all that view them. The Game was to see if total Free Choice would work. Not just for you, but for all that is. From the start, it looked as though it was not yet time for a Game of such experiment, as the Game brought out the very worst in some. Even with all the setbacks and time running short on the Gameboard you still managed to find your way Home. You truly are a grand expression of spirit indeed! We are proud to call you family.

It is with the greatest of love and honor that we ask you to treat each other with respect, nurture one another and play well together ... the Group

"Awakening the Master Healer" exercise at the Scepter of Self Love seminar, Denver, Colorado

Chicago was warm that day, and even though everyone was in a hurry to get off the stuffy plane, I stayed behind to catch a glimpse of this mysterious gentleman. I simply had to put my mind at rest. I was the last one off the plane and he was no where to be found. I looked around the terminal but I never did see him again, at least not in physical form. Did I imagine this? Or was something much higher happening here? In any event, the Gentleman with the great white beard is back. With his finger lovingly held high in the air, the Group is silent in anticipation of his action as he majestically lowers his hand. This simple but important act signifies that **now** is the time to release this information and make space, once again, for the Master Healers. By the way, if you happen to see him around, please tell him hello from the one they call the Keeper.

Chapter 4

Relationships

Getting Along in the New Energy

Relationships

I re-member reading a book once that really resonated with me. I re-member saying to myself "I wish I could hear spirit that clearly." It was during a time when many of the people in my life were discovering their spiritual names and the names of their guides. I thought this was very important because it meant that you had a good connection. I know this may sound odd, but I am a bit of a skeptic and am always looking for some kind of verification. Even now there are still days when I wake up and wonder if perhaps I'm just making it all up. I just want to be assured that this is real. After all, if you look back at my life thus far you will see that I made a huge shift from General Contractor to Lightworker in less than a year. A short time ago I would have laughed if you had told me that I would be helping people move into their path by channeling information from the other side. I guess the joke is on me. Still, somewhere deep in my heart I can say that I always "knew" this would happen.

One thing I am finding is that I am receiving verification all the time. As usual, the message is always for me as well. The Group has this strange gift of speaking to many people at one time and often even conveying different messages with the same words. The Group has told us there are Nine members, of which they consider me one. They all have very different "personalities" and fields of expertise. My field of expertise is the human condition and my job with them is as a translator. Their message is one of empowerment. They want us to know that *we* are the powerful ones and that they are here to help us re-member and use that power. The many times I have asked for their names and origins have left me with the same answer. Basically they say, "it's none of my business". They have told me that they have never been incarnated here on the Gameboard Earth, as they call it.

It was not long before I understood that I am not the only one who channels the Group. They come to many others in different forms. Always, there is the unmistakable love in their message. Part of the reason I have been encouraged to do this work is to help people re-member that they also have this same connection with Spirit.

The Group has asked me to find other words to describe the process we call channeling. They say that this word has much misplaced energy around it that limits our understanding of the process. Within our workshops and seminars, our efforts are now directed toward helping people develop their own "conscious connections." Encouraging this connection within ourselves helps each of us to facilitate the evolution of mankind.

The Group says this "channeling" of spiritual entities is temporary and only a warm up for what is to come. The evolutionary process will provide us with the tools to connect with our own and each other's higher selves. This, they say, will lead us to the ultimate in information and communication. Can you imagine a world where we can all understand each other's thoughts and feelings? This will certainly change the basis for many relationships here on the planet. Knowing that everyone has access to each other's thoughts may take some getting used to. Yet this will be a world where there will be no room for things such as war. After some awkward adjustments, this will provide the basis for Heaven here on Earth. The Group calls this "Earth's next step."

Much shifting is occurring on the planet at this time as we prepare for this next step. It feels like the cauldron is beginning to heat up and someone is stirring the pot. A lot of people have felt this stirring in the area of relationships. As the evolution of mankind continues, so does the basis of our relationships. It may seem strange asking for information from entities that have never played our Game. I am glad I did because their answers have surprised even me.

The Group:

We welcome this chance to address the gathering of masters within this family. As the Earth continues shifting dimensions, many on the planet are rapidly moving into the areas that have been requested. Let it be known that these personal shifts affect only those areas in which the higher self is in full agreement. Of equal importance is that all these shifts are always initiated for the highest good of all concerned. This is a time on the planet for repositioning. As this unfolds, each of you will assess your present situation and ask if it still serves your highest good within the setting of your new contracts. Although this process is healthy and natural, for many of you it brings pain as you have spent so much energy holding a vision of something you refer to as "normal." With more information comes a new perspective and new visions. The Keeper often uses an expression that states: "Everything I ever gave up has claw marks all over it." Allowing these new perspectives to manifest themselves in your everyday life now becomes the challenge. Viewing your relationships without judgment is difficult for most because of the attachment to outcome to which most of you so desperately cling. **Release** *and* **allow** *are two words that will serve you well in the times to come.*

The object of the Game was for the Infinite Creator to be able to observe himself - herself. This is accomplished on the Gameboard of Free Choice through the eyes of other players. This you have called relationships. These relationships can be a source of both joy and pain, yet they always reflect a part of you that is magnificent.

CENTERING YOUR ENERGY

A viewpoint that consistently produces results other than what you wish is that you "need" others of similar vibration to validate yourself and make yourself whole. This is an effect of polarity on the Gameboard and is nothing more than illusion. You are a whole entity unto yourself alone. When you speak to God is anyone with you? When you sit in your chair of responsibility are their two seats or one? When you retreat to your private place to re-create yourself, is someone beside you? These are sacred spaces within you that have no polarity. Here, there is no need for another to validate you. These are safe havens for you to re-member your true essence. This essence is yours alone. Have the courage to place yourself first in all matters. For only then is it possible to walk in full unison with another. There is but one true relationship: That is the relationship of You with Yourself.

THE TRUE MEANING OF LOVE

In your relationships with others you have asked many times about which way to turn... "give me a sign", or "Spirit bring me a mate". As you are aware, we do not act in such a manner for several reasons. To do so would be against the highest interest of the planet. We will share with you glimpses into potential realities to renew your hope. If you wander too far, we will lift the veil and touch your heart so that you may re-member Home. We will be there to light every step of your journey, for we hold you in the highest of honor and love. We will encourage and instruct you to go inside and retrieve the answers for yourself. In doing this you will learn the much higher lessons of self-empowerment and discernment. You already know the truth within, this is only an exercise in re-membering. Most of you look to your heads for the answer instead of trusting your own hearts. Learning to access that truth within is the challenge you face. This is illustrated within the field of polarity as learning to access your feminine side.

You have also asked about how things exist for us on this side. It is now appropriate to reveal some of these truths to you. Your advancement into the next stage of evolution is drawing you closer to our own vibration. You are literally getting closer to heaven. This transition is also a move out of polarity into unity. Unity consciousness is not fully supported on the Gameboard at this time. This is changing. Unconditional love is what we would refer to as Universal expression of the Love energy. In polarity, the concept of unconditional love is not fully supported. This is neither right nor wrong. It is simply a statement of where things are at the moment. Your relationships at this time are based on conditional love. "I will love you as long as you love me," you say to one another, expressing the conditions as well as the love. This has worked well for you and has provided a basis for the energy of relationships in the vibrations of Earth up to this point. Now that the vibrational rate of the planet is rising, however, the collective vibration of humanity is also rising. New possibilities are surfacing that will finally support these higher expressions of the energy called love. Incorporating the application of unconditional love into the relationships now in place will make it possible for you to make the transition to the higher vibrational levels.

RELATIONSHIPS IN SPIRIT

We, in spirit, carry both male and female energies at all times. And so do you. Once again, it is the polarity of your environment that has you dividing yourself into one or the other of the polarities, or sexes, for purpose of expression on the Gameboard. It is this same polarity that causes you to view us as either male or female. On this side of the veil, we exist as whole beings and we clearly see that all is connected. Without the veil, we experience this connection in every movement. With this connection we are able to easily carry both the male and female within us at all times. Your ancients have termed this the Yin and the Yang. On this side, we do not use two words to describe that which is one. On Earth you look at the

Yin and Yang and you see two parts coming together to make a whole circle. From this side of the veil this appears as a dance of two complete circles complementing each other.

We are each complete unto ourselves and therefore do not play the Game of "twosome" that is prevalent on your Gameboard. Please understand that the Game as you have defined it has been perfect for the state in which you have been living, and we honor the Game as you have written it. In comparing our side of the veil to yours, you can see that the same Game would not work. Perhaps you can now see that to ask us which way to turn in these matters would not be appropriate, for this is **your** Game and the rules are of **your** making. The love we have for each other here is so great that it would be difficult to describe in this medium. Suffice it to say that relationships in our dimension could be best illustrated by viewing the actions of one of the dolphin families on Earth: all members move and act together in complete support and love for the whole.

The larger Game of Hide and Seek that you are playing has a goal of re-creating Heaven on Earth. Here is the key. It is only possible to re-create home by finding balance within polarity. Once balance is found, the Game moves to a higher vibration and you experience the grand shift. After the shift, polarity is no longer needed. Balance is what you have ventured to attain every moment of your life thus far. It is in this search for balance that you have devised this act of partnering. Partnering has served you well and it is highly honored. It has allowed you to meld into one another, sharing vibrations and producing overall balance. We on this side of the veil are in awe of your ingenuity at overcoming the effects of polarity. To hold another close enough to see your true self in their eyes takes true courage and honesty. As time progressed, this Game of twosome evolved into many sub Games. Some of these were about power and control and some were about self worth. Others were used to obtain material assets and personal comfort. Some bounced these relationships off because of judgments they carried about themselves.

RELATIONSHIPS AS MIRRORS

The basis of relationships on the Gameboard is to mirror your own magnificence and re-mind you of the part of God within you. Some relationships have held difficult lessons for you. This is as it should be, for to see the God within you, your own energy must be centered as it is only visible through eyes free of judgment. There are times that you call mirrors into your lives that reflect parts of you that you do not like. Use these mirrors to openly see the reflection they offer. As you begin to shift the reflection within the mirror you will also observe the mirror shifting as well. You have been told that you cannot change another person. We tell you that in this fashion it is possible to change another by changing the way you react to them.

Once the sight is gained into oneself and the lesson is realized, a new relationship will emerge. It is now a new person being reflected in the mirror. For some, this will mark the end of a relationship. Some relationships will make the transition and move to the next level. There is no judgment one way or the other as to which is best. The only thing we wish to point out is that in most cases you have spent a great deal of energy placing this mirror in your field. Do not remove it before you have seen the reflection as that will only cause you to seek another mirror of similar aspect to gaze upon. Here we ask you to use discernment in making these choices. We also re-mind you that it is not just the person being reflected who must change. Once a change is effected in one party, the mirror must also be willing to hold the new reflection. A mirror unwilling to cast a new reflection is forever stuck in a limited relationship. We will tell you that the mirrors that welcome the opportunity to change their reflection are the most likely to shift to the *higher levels. This often requires forgiveness and understanding. This illustrates the use of unconditional love and the Unity aspect of relationships in the new energy.*

Many in the higher vibrations have found difficulty in creating relationships that feed them. They see themselves as lonely and unable to connect. We re-mind you often that you are much more powerful than you understand. Your thoughts create your reality. Relationships are no different. When you send out a call to the Universe for another to share your life the Universe returns the exact same vibration that is sent out. The Universe has only one answer to all of your requests: "And so it is. . ." The only restrictions placed on you within the field of relationships are those that you place there yourselves. An honest look at the results of your requests will help you understand the vibrations that you sent out. Some look at their bodies and think quietly to themselves "who would want me?" Others want desperately to have another in their life but are unwilling to be vulnerable. We tell you that these thoughts are the first ones to precede you into relationships. Often these thoughts keep people from ever finding you even though they heard the original call. Often they hear "Come here. . . come here" and as they approach they hear "Not too close!" This also attracts people who are emotionally or otherwise unavailable. After all, this is a direct response to your own request to the Universe. We tell you now that there are direct matches for everyone on the planet. Release the comfort of your loneliness and the judgment long enough for them to find you. Release your own judgments about yourself and allow yourself to be totally open and vulnerable. Do this and watch the doors fly open. Allow yourself to be an accurate and forgiving mirror and also to look into the mirror before you with anticipation of the greatest reflection of God.

Most of your partnering is done by deliberate contract. In an attempt to balance, you attract those into your lives who have the attributes you need to attain this balance. Often this is a contract you made prior to entering. It is by this mechanism that some have contracted to repel partnership in their lives. These contracts have a specific purpose that you have planned out in advance. Many of these contracts are based on raising children. Sometimes the contract is accomplished and there is a push to move on. Other times the contract is complete and still it is appropriate to stay. If you will remove judgment, it is easy to see that all of these contracts are appropriate. Many contracts made in the old energy, however, will not survive the shift to the new energy. This we are seeing now.

Also, if at any time you choose to turn from the relationship without fulfilling a contract, you become predisposed to attract another with similar attributes back into your field. This is why some seem to re-create the same scenario time and again. It is because they left before they were finished. In all cases they were attempting to walk away from attributes within themselves. This thing you call relationships is the grand mirror indeed. It is a gift you have given yourself of the highest order. Look into this mirror closely and walk fearlessly into the lessons you have placed there for yourselves.

REVIEWING YOUR RELATIONSHIPS

As this time of Merlia continues to grow stronger there will be many who will examine their motivation regarding the relationships they are now in. They will ask a multitude of questions such as, "Is this relationship complete?" "Was this a contract I made in the old energy and is it still in my best interest?" "Am I here only because I am comfortable or does this reflect my highest potential?" Understand first that all of these relationships were appropriate and honored. There is no judgment about motivations, except those you have about yourself. Also understand that you humans create layers of complexity in relationships with your early childhood judgments. As children you search for that which can be framed and labeled as "normal." When you find this image you place it on the wall and use it as a measuring stick to gauge all that enters your environment. It would serve you well to examine and purposely change these images if they are not in your highest interest. It is these pictures that direct much of your life and very often your relationships. As you advance so should these images. Do you hold all of the same truths that served you as a child or have you moved into higher truths as you moved to higher vibrations?

MAKING AND BREAKING RELATIONSHIPS

As the vibrations of the planet increase, the veil will thin and you will be re-membering who you are. Until such time as awakenings begin happening to the masses, it will appear to those around you that you have changed. This change will spark many shifts. People will suddenly appear in your life carrying the vibrations of Home. These will be the members of your original spiritual family. Many people in long-term relationships will suddenly wake up and find they are living as strangers. Groups will form instantly to support each other as spiritual families reunite. Long time friendships may quietly drift off and become insignificant. The value of these relationships is not to be discounted, yet many will not recover from the shift. Some will end in anger and fear. Some will find love within themselves and partner for the first time. Some will part in love. Some will rebuild from the ashes and start anew. These are only possibilities and they await your direction. For it is you that will decide the next step of your relationships.

LOVE FIRST YOURSELF

And what now of love? Here is the answer to all relationships. **Love first yourself.** As we have explained before Love is the mother of all energy, and therefore contained in all things before you. Much of what you term romantic love is an expression of the Kundalini energy that is often misunderstood. This energy and how it relates to sex will be an important topic for another time. To express love in all of its forms is your divine expression, yet there is one that must come before all others. That is the matter of the inner heart, the love of self. You on the Gameboard of Earth have volunteered to walk in three- dimensional bodies, behind veils that limit your sight, for the good of all that is. For this you are loved beyond your understanding, yet you refuse to love yourself as the part of God that you are. Many announce quietly to their inner selves that they will love themselves only when external parts of their existence are in order. "All will be fine when I lose that weight, get that job, have that family, buy that house." Such is the reason they feel emptiness in their lives. They have taken the power from within, where all is real, and placed it on the outside, where all is illusion. They have relinquished their power by failing to understand the true nature of the Game.

Love first yourself. There is no greater expression of love than the love within. Look into your lover's eye and you are taken with a feeling. You are elated and empowered because of the way you feel about yourself. All is attainable in this state. We tell you now it is because your lover allows you to see yourself in love through the mirror of their eyes, and this you will accept. This is a way in which you will allow yourself to love yourself. It is appropriate and honored because it is a divine expression of love. This is God loving God. As a mother looks into the eyes of her baby she senses the same truth and feels good about herself. Some will view this as self centered and selfish, yet nothing could be further from the truth. For it is only when you allow yourself to become whole that you are able to give true love to those around you. Until you truly love yourself there is nothing to give to others, and therefore the energy will not circulate. There are those of you who consider yourself lonely and see yourself as having difficulty attracting a partner. To you we say look first to your true feelings about yourself as those are the vibrations that are sent out to the universe to manifest. *Another will never interrupt to keep you from loneliness. Only you can accomplish that.* The need to focus attention on another to keep from feeling lonely will only shift the energy from loneliness to dependency.

Unconditional Love

And what of unconditional love? *This is the natural law of attraction at its best. Every act of creation in the universe has motivation. Look to the motivation in order to observe the true flow of energy.* To give love freely, openly and selflessly sends a message to the universe that automatically returns all of like vibration. Unconditional love illustrates an understanding of this law. *It also represents a shift from polarity consciousness to unity consciousness. This is a purposeful step into the next level of humanity. We illustrate with previously channeled information, "For as ye sow, so shall ye reap." "Do unto others as you would have them do unto you."* These are illustrations of the law of attraction. *What is omitted here is that the vibrations are returned to you amplified many times over what was broadcast.* Balance the conditional and unconditional love in your relationships. *Unconditional love is a discipline that requires removing all judgment from your field. As we have said prior, all judgment is traceable back to judgments you have about yourself.* Once again you see that all relates to your feelings about yourself. Love first yourself and all things will come into focus. *If you find you are unable to love yourself first, then open to the love we have for you, for we are here to love you until you learn to love yourself. . . Our motivation is simple.*

Please re-member Unity on the Gameboard and treat each other with respect, nurture one another, and play well together. . . the Group

I am re-minded often by these loving beings that this is a Grand Game of Hide and Seek. They say that we are pieces of God hiding behind veils that limit our sight. And the Game is to see if we can recognize one another. We ourselves penned the rules by which we play this Game, although at all times we have Free Choice. As we evolve, many of the rules that seemed to work in the past now appear to be keeping us from attaining our goals. These are exciting times as these changes happen before our eyes. It is an age of new paradigms and a time for shifting into position for the next phase of the Game.

If your relationships are not all that you choose for yourself, take a good look at those rules and pictures hanging on the wall. They may reveal some answers. Perhaps they are from a time long ago and are in need of updating. This takes courage as most of us have built our very existence around these rules.

In San Diego, where we live, we often go camping at an Indian reservation. The favorite sport there is to get in the raging river on an inner tube and see how long you can survive. It gets a little scary sometimes, but generally it's a lot of relaxing fun. One day, my dare devil son, Austin, decided to take his inner tube down a 12 foot waterfall. I held my breath and watched from below, preparing to pick up any pieces that floated my way. On his first descent he got beat up a little but immediately started back up the path for another ride down. Glancing over his shoulder at me with my mouth hanging open, he asked if I wanted to join him. There were young women standing next to me and he knew I was macho enough to be manipulated in that fashion. So I climbed up the trail with my inner tube and tried to act like I knew what I was doing. I proceeded down the falls and got beat up a lot worse than he did. This had something to do with a law of physics concerning the weight of bodies in motion, of which I did not have a complete understanding at the time. Now picture this: I am now at the bottom of the waterfall

Illustration: Phyllis Brooks

trying to locate my swim trunks, when I look up and see Austin going over the waterfall without his inner tube! Before I have an opportunity to scream my tidbits of wisdom at him, he is standing unscathed before me saying: "That's co-ool! Try it dad, it's easy." He must think I'm stupid. (Okay, so he's right.) A few moments later, after tightening the drawstring to my trunks, I am on my way down again, only minus the inner tube this time.

To my surprise, I actually made it down the waterfall effortlessly. Apparently the rocks making up the falls were covered in thick, slippery algae that had the effect of lubricating and cushioning the ride. This could not be seen unless you were actually on your way down and in the channel that the water had cut out of the rocks. The inner tube made it impossible to get in the natural flow of water. The lesson I learned was this: Hanging on to 'the inner tube' of our belief systems often keeps us from finding the flow and the path of least resistance.

I spent the rest of that day with my son going down the waterfall as onlookers gasped and tried desperately to hang on to their own inner tubes. They were tossed and thrown, bruised and beaten, as they tried to "conquer" the falls while hanging on to their inner tubes, which they believed held their security. They brought with them their ideas about the rules of the Game. My son decided to change his and follow the natural flow.

Thanks for the lesson, Austin.

Chapter 5

Sex

Understanding the seeds of your past

Natalia Kavtaradre

Sex

One of the important topics the Group felt it was very important to address was sex. They told me a long time ago that what we perceive as sex has a much deeper meaning than most of us are aware. They said that the "attachments" we have placed on sex have made it appear to be something other than it really is. To us, it is a way in which we can express love in its deepest form. They say that for us, especially in our society, sex has become a form of power that we often use on each other. What do they have to say about it's true meaning? The Group says it's a way to experience Home.

The Group:

It is with great pleasure that we gather here to speak with this growing assemblage. The work you do here may appear to be limited to your individual selves. This is not the case and we tell you now that it reaches far beyond your scope of understanding. You are most honored for the work you do within. This is the first task of all that choose the name Lightworker. It is from this perspective that you can most easily change all that surrounds you. To change the building blocks that assemble the universe, you can either start anew and carry each block into place with your hands, or you can plant the seeds within and let your co-creations materialize naturally. Truly, we stand in honor of you and your choice to move forward on this path. The time is right for you and each one of you is needed at this time. Your work here reaches far beyond your current scope of comprehension and we thank you for all that you do.

As we have spoken before, there are parts of the Game you have been playing that continue in a direction that does not serve your highest good. It is time to plant the seeds that will allow a natural correction of this course. Trust in your own powers of discernment, voice your intent that all information will be carefully screened for *your highest good, and so it will be. Know also that there are no seeds that can be planted without your approval. We encourage you to take your power, and give it up to no one including us, for you are the rightful heir to your own destiny.*

HISTORY OF THE GRAND GAME

As we view the Game in progress, we are astounded at the ingenuity and creativity you have applied time and again to overcome the adversities that faced you. The Game, as planned, did not include much of this adversity. As a result, there were times when you found yourself facing situations and opponents for which you were not fully equipped. The Game then took several unexpected turns, and sometimes was even thought to be lost. It was at times such as these that you found your greatest strength lay within, and you called upon it. This is when you discovered that you could plant your own seeds and create your own reality. The seeds you planted and nurtured within yourselves became the basis for the new paradigm you are now calling into existence. This is the return to the garden you were promised in the beginning. You have made it so. We stand in awe of you and your greatness.

The time has come now to understand a little of your own history. This knowledge will help you to traverse the path Home that now lies before you. Early on, when the Game was still new, there was a faction from a neighboring planet that attempted to control you as a race of beings. At the time you called them "gods" because you perceived them as superior. They allowed you to give them your power and call them "gods." Their plans included keeping you from knowing your true heritage and therefore your own power. Since you had given them your power, they planted many seeds within you to control you. To this day some of those seeds still have their roots deep within your being and occasionally re-surface. Our intent is to make you fully aware of these seeds, and thus allow you more freedom of choice in your thoughts and actions.

The captors' original intent was to create a race of subservient beings. These seeds that they planted have re-sprouted several times in your own history without your direct knowledge. As a result, you have continued unconsciously to perpetuate this pattern. There were times in your history when you allowed what you call slavery to be an acceptable behavior as part of your Game. This, along with racism, is a direct result of these seeds re-surfacing. This, as you have discovered, is not in your highest good. Understanding the reason for your tendencies will help you to determine your future path. You have propensities in several areas because of seeds that were planted then.

One of the ways these beings chose to control you was to plant seeds within you that told you that sex was inappropriate. Your captors did not allow sex, because it would have provided you with the inner knowledge of your own powers of creation. Thus, you were told that "the gods did not approve of sex." At this time in your history, when most of your powers had all been stripped away, this was the one area of your power that your captors could not fully hide from you. This was your power of **pro-creation**. There was a biological urge that simply could not be hidden from you completely. The seeds they planted at that time were mental seeds that told you that sex was wrong and something you needed to learn to control. This has had a far-reaching effect that lingers still within humanity today. These seeds have re-surfaced throughout history in your societal views and this is the main reason that sex has never been considered a spiritual path. We tell you now it is a direct reminder of Home.

After a time, you began to discover some of your own powers through pro-creation. As each of you began re-claiming your own power, you became "spoiled" in your captors' eyes, so they cast you out of their midst. Because they were concerned about influencing the others in captivity they reluctantly granted you your freedom. This was later referenced in your own writings as being expelled from the garden. Many of the stories from that age have surfaced in the earliest writings of what you call the Old Testament. Over time, more and more of you began to rediscover your powers through the use of sex. Finally, the idea of a subservient race was abandoned all together. This was a grand time for you as you began to use your own powers for the first time. The renegades who attempted this control had also reunited with their own elders during this time. It was a time of great healing for them and for you, and much was gained from this universal lesson. In the later years the renegades and their elders did much to help you become self sufficient before they left the planet altogether. They have now returned to aid in the transition before you.

In their efforts to control you, this race physically altered your DNA and blended it with their own. As a result of this blending they are now considered one of your six parental, or root, races. They are here at this time and are playing a large role with the current evolution now in progress. The work they are undertaking includes planting seeds to re-pattern your DNA to accept the signals that have been outside of your range of reception ever since they first altered it.

THE RESURGENCE OF POWER

There was a time not long ago when many of you who started this Game decided to inhabit the Gameboard en masse. The idea was to all be there at the same time and carry so much light energy that it would push the pendulum of light and dark so far into the light that the Game would be forever changed. This was the first time a critical mass was reached and it was successful. This resurgence of light began during the time you have labeled the Baby Boom era. This is the reason that the 1960s and 70s carried with it the flavor of a sexual revolution. This was a replay of the time when you regained your freedom from your captors through the rediscovery of procreation. The pendulum that swung very hard to one side during that time is now finding its balance and coming to rest somewhere in the middle.

These seeds of power surfaced once again at a crucial time when a critical mass (i.e., the Baby Boom Generation) was reached. The difference here is that you planted these seeds yourselves. When you allow such things to take place, you become the Creators of your own destiny. It is your true heritage.

Seeds in a garden lie dormant until they are activated by a combination of water, temperature, and fertile soil. Many seeds you carry within never got activated at all. Still, they are there waiting for the right conditions to trigger their growth. This is the case with the seeds of power now sprouting within you. The vibrational realignment of the Earth and the higher vibrations in which you now reside are creating the ideal conditions for these seeds to sprout. Some simple triggers for this process are being seen in the master numbers that you are now calling into your field. These simple yet effective triggers set up the physical conditions that encourage the sprouting of the seeds.

Because of your history of being controlled, you have a dichotomy with the issues of control and power which you express through sex: Either you give your power to others readily, or you assume a posture of taking power from another. Such is the case when you use sex for control and power. In your field this is more prevalent than you may be aware. Understand that this is nothing more than a sprouting of seeds that were planted during the attempted control of your race. Recognize this for what it is, and you will quickly gain the tools to move past it.

There have been many truths uncovered pointing to the spirituality of sex. One in particular was made by a splinter group of a

religious order that received information that led to the development
of what you call Tantra. This was clear information that was brought
forth at a time when the pendulum was poised for a violent swing to
one side. With this information in place, it eased the necessity for
the pendulum to swing so far, and more of a balance was maintained.

The biggest challenge you face is accepting your joy when
faced with it. The attitude of humanity on the Gameboard convinces
you that you are not worthy of joy in your life. Many of you have
therefore built protection in layers to keep you from experiencing the
joy. Even though this attitude is predominantly traceable to events
within your current lifetime, it flourishes easily because of the fertile
ground created by your experiences of long ago. Time and again
your individual lack of self worth has stopped you from manifesting
your co-creations and experiencing joy.

When you planned the Game you placed things in your path
that were very important to you. To these things you attached a special feeling of joy so that your biology would have no choice but to
readily accept them as truth. What happens instead is that you feel
the joy but negate the experience because you believe yourselves to
be unworthy of such joy. There is great humor about this on our side
of the veil. It is difficult for us to understand that the veil is so very
complete that you cannot see your own magnificence and your own
heritage. We are here to help you understand the importance of this
joy in your life. One of the ways this joy can be experienced is
through the honest act of sex. Joy and passion are the road signs to
your path Home that you placed upon your own path. Heed these
signs and take them unto you. Your joy and passion will heal you
and lead you back home. Recognize these as the crystals that you
have placed on your own path to guide your way Home.

Biological life, the Game, on planet Earth is based around finding balance in all things. The natural flow of energy throughout the

Universe is one of constantly seeking balance. Master that balance, and you will claim your master status in all areas. Achieving balance is the central theme of all that you call nature. It is prevalent in all things around you, yet when it comes to applying it to yourselves you often toss it aside. We say to you that this is a key to the Game. Find balance in all things. "Up" has no direction without "Down." "Right" is not correct in the absence of "Wrong." If it were not for the darkness, light could not be seen. It is the balance that gives value to the whole. It is for this reason that the blending of your races on Earth is a most honored biological step in this evolutionary process. This is now well underway. Look for the blend and balance in all things.

KUNDALINI IS YOUR OWN ENERGY

So often one looks deep into the eyes of another and sees oneself reflected as the fully empowered being they have been striving to become. This is the Kundalini energy surfacing and can easily be confused and misdirected. Kundalini energy is one of the most powerful energies that you carry within your body. You have called it sexual energy because it arouses all of your senses without exception. What is happening is that all of the chakra centers open to allow the flow of this energy through the body. When this occurs for the first time it is naturally foreign to you. When it flows in through your being you are momentarily at the height of your power and this can be frightening. Many of you mistakenly attribute this energy to your partner instead of accepting it as your own. This is a natural awakening and will be happening more frequently as you raise your vibrations. As you look into another's eyes and feel this energy, what you are really seeing is the reflection of the God within yourself. It is a very powerful event and you walk away forever changed.

There are many of you now who are experiencing this Kundalini energy. Take it without shame and thank those who *mirrored it for you. Experience this joy. Many of these contacts are instigated by contracts that were made long ago. They serve to remind you of your powers and the direction of your path. The exhilaration that accompanies this experience is often mistakenly equated to falling in love and often confuses those whom it touches. Take this in your stride, and know that as you become accustomed to carrying this energy, it will become second nature to you and bring you more joy than you can imagine. With practice you will be able to look in the mirror and see the God within your own eyes and accept this joy readily. This will clearly mark the height of your empowerment. This is the true* **you**.

We ask you to take these words and thoughts presented to you and apply your own discernment to them. If the message rings true in your heart take it as your own. If there is anything less then leave it without judgment, as it was meant for others. It has been our great honor to present these messages to you.

It is with the utmost love that we ask you to treat each other with respect, nurture one another, and play well together… the Group

I noticed that the Group had several opportunities to name names and point fingers in this message. Instead, they specifically chose to leave it open. It is important for them to remain open. I don't yet understand why, but I have experienced this attitude with them before. They are reluctant to put out information that can in any way take away our own power. We humans do love to give away our power. I specifically got the feeling that the only reason we were talking about this was to explain the seeds that were planted, and what to do about them.

While writing I got pictures that were difficult to describe. As they spoke about our "captors" I saw pictures that were more suggestive of us willingly following a leader, rather than being led in handcuffs. This attempted takeover was very subtle. I got the impression that we were rewarded for our allegiance, so on some level it was a choice of ours to follow these "gods." The Group also insisted that I use a small 'g' when referring to them. The unfair part was that these captors took advantage of the new Game of Free Choice and tried to control us in ways for which we were unprepared. This was a renegade group acting on its own, and there was even dissension on their home planet about their aggression. Seems that we all learned some great lessons here.

One thing I noticed that was not addressed in this message was the appropriateness of what happened. The actions of the renegades were acceptable on some level. Actually, I got that this might have even been sanctioned. By whom or what I have no idea at this point, but I did want to mention that I felt this very strongly. If you read "Conversations with God" by Neale Donald Walsch, you may remember the point at which God said that he did not care what Games we played because we could never really get hurt. This was the same feeling that I got; a feeling that no matter what happened we were never in any real danger. This is a wonderful feeling and I must say that I felt it a lot during this writing. This is the same feeling I get when they refer to life here on Earth as "the Grand Game of Hide and Seek."

I should also note here that during editing I went back to remove the part about balance, in which they went out of their way to explain that the blending of races is a reflection of the natural energy of the Universe and is "well underway." I thought it was interesting, but out of context, and I wanted to move it to the chapter about biology. The Group had other ideas and that particular portion remains intact here.

Sex on the Other Side of the Veil

Since I originally wrote the chapter on Sex, the Group has

provided me with further clarification on sex as it relates to their existence. I have always said that the Group is on my shoulder all of the time. When I first started channeling, Barbara naturally wanted to know if they were on my shoulder **all** of the time? The answer came in the form of that wonderful laughter that I love to hear. Because, of course, they are indeed with us **all** of the time. And what's more, they love it when we make love.

This naturally raises the question: Is there sex on the **other** side of the veil? I have been told by some very big names in this field that there's no such thing as sex on the other side. Moreover, sexual abstinence is mandated in many religions as a path to higher spirituality. Since I have been receiving information from the Group, I've had to let go of some of my beliefs to make room for the information they offer. This particular subject proved to be no exception.

The Group has told me many times that although they appear to me in the form of male and female, they do this only to make it easier for me to relate to them. One member of the Group I've spoken about is the scientist. This energetic, almost hyperactive personality is constantly changing sexes and shifting between polarities. It has been difficult to write about this entity because in one sentence I use the word "he" and in the next I find myself using the word "she". He/she says that they switch polarities to illustrate the physical characteristics of the unity consciousness.

The way the Group talks about it, polarity is an important part of the mechanics of the Gameboard. The whole point of the Game was for God to see herself/himself. (Re-member, an infinite being is not capable of seeing itself.) Polarity was necessary to provide the contrast required to see the parts of God. The effects of polarity also created some necessary illusions. Looking through a field of polarity we perceive ourselves as separate from one another. In truth, we are all one. Even though we inhabit separate bubbles of biology while here on the Gameboard, we are always energetically interconnected on every level. However, in order to enter the Gameboard of polarity, it was necessary to split from the whole to assume a finite expression. This could only be achieved by assuming one polarity (gender) over the other.

In each incarnation we choose the sex for each phase of the game. The two sexes have different attributes to allow for lessons to be fulfilled. In some incarnations we chose a gender for a specific purpose, only to find later on that our hearts were still firmly attached to our previous gender. The Group explains that this is the basis for what we call homosexuality and bisexuality. These are natural conditions that have existed since the Game began. Although many carry judgments about this it is a part of the expression of love that is the spark of God carried within each one of us.

So the question still remains: Is there sex, as we know it, on the other side of the veil? The Group says that when we experience each other in intimacy we are actually experiencing glimpses of Home. When two physical beings connect in this manner they momentarily balance each other's polarity. This instantly connects us to the vibrations of Home. Since there is no polarity beyond the veil, it naturally follows that this ecstatic connection is a permanent part of life on the other side of the veil. There is no act of sex, as such, on the other side of the veil. Rather, it is a constant vibration. Here on the Gameboard, we get to experience this vibration —if we are lucky—as occasional orgasm. On the other side it is constant. This is the reason that Spirit loves to see us connecting in an expression of Love. When used appropriately, sex is really just our way of re-membering the vibrations of Home.

Chapter 6

Tools for Walking With Spirit

Balancing your ego for the journey home

Barbara and Steve finding balance in Aarhus, Denmark

Balance

Three steps forward, one step back. This is often our perception of growth here in the human state. There are even times when it feels like one step forward and two steps back. We look at this and want to find ways of speeding up the process. In asking spirit for help we often forget that they have a very different perspective on our situation. We ask for what we think is in our best interest, instead of simply asking our higher selves to lead us to our highest good. Filtering everything through our own higher selves would resolve most of our issues very quickly. The question then becomes, why don't we all do it all the time? The simple answer is, there are forces at work that we do not yet fully understand. These forces are things we have created ourselves in most instances, and they served us well for much of the Game thus far. The Group says that it is now time to disclose more information about these forces and how they have kept us separate from our higher selves. Learning to walk each step of your lives holding hands with your higher self is a process of conscious living that the Group calls Re-membering.

The Group:

It is with great honor that we present this information. Those of you who feel drawn to our messages do so because you recognize the familiar vibrations of Home. This is a time to re-connect with those vibrations. Through your intent to move forward you have called this family back together. We offer you this information for your own discernment and ask you to filter everything through your own higher selves. Your higher self is the piece of you that you call the God within. It is a much larger part of you than you presently see. This is your true identity and your heritage. We are here to help you re-member this, and to walk in the power of your own vibration. We are here to help you re-member who you are. We are here to help you understand that the greatest power of all lies within yourself. Expressing that power in your present state is a re-membering process and is the greatest expression of the God within each of you. This is what we mean when we speak of creating Heaven on Earth. This Re-membering is the ultimate outcome of the Game now being played on the Gameboard of Free Choice.

FROM BODIES OF PURE LIGHT

When humankind first began the Game, you occupied an ethereal form composed mostly of light. This served you well, as you were equipped to channel the vibrations of light into the Earth. As the Earth cooled her vibrations slowed and it became harder for you to interact and accomplish your objective. It was then necessary to increase your density to match that of the Mother. This was when you began the process of ingesting foods of the Earth and also taking in less of the light energy. This is a natural evolutionary process. At the time there was much excitement around this shift as the players began a new dimension of the Game. This grounding process you call eating is highly honored, for it allows you to transform the ethereal into a tangible form, which is your highest objective. Today, this is what you refer to as bringing light to the Earth. This is the ultimate objective of humankind. In these writings we often refer to it as bringing Heaven to Earth.

During this time of adjustment you developed parts of your emotional body that helped you survive. As you gained density you became vulnerable to certain aspects of the Gameboard. At this point it was necessary to develop an internal focus of survival. Survival has come first and foremost in all of your thought processes since that time. As a result, the need for survival became deeply embedded in your emotional body and has since permeated all areas of your thinking. It has become so deeply ingrained that you now carry it in every cell of your biology. This instinct, as you now call it, is integrated at such a deep level that most of you see it as a natural part of yourself and readily accept the influence it carries.

It is this focus on survival that is the origin of many of the fears you now encounter. These are the unseen forces within your field that often cause you to take unnecessary detours on your path. They often impede your progress. It is important to understand that these forces carry no energy of their own save that which you give them. Fear them not, for to do so only gives them power they do not deserve. All of these energies are within you and, as such, can be redirected to a more appropriate alignment. It is humorous to us that you refer to these as "negative energies," for when you first received them they were highly revered. It is not the energies that have changed; it is you that have come full circle. You are finding your way Home again.

THE EGO - A TOOL FOR SURVIVAL

You are at a time in your evolution when it would be helpful to rebalance these energies. Upon tracing the energy back to its origin, we see that this manifests in your life as primary fear. On some level, all fear is traceable back to this original fear: the fear of dying. This survival mechanism is rooted so deeply that it very quietly alters your actions. To balance this primary fear you have developed what you call the ego. This is the part of your emotional body that allows you to be in the dense vibrations of biology. If fed too much energy, the ego takes the posture that the biology is all there is and negates the spiritual side altogether. Above all, it has the job of ensuring survival, thus, it will stop at nothing to complete its prime objective. This is part of the illusion you walk with behind the veil. Primary fear is actually the great motivator. You have found relief by balancing this energy with an ego that gives you a fuller sense of self. This is as it should be. Yet it is important to understand the universal truth that we are all one and not separate in any way. The ego is *simply a tool and nothing more. Use this tool in the application it was designed for and it will be invaluable. Use it inappropriately and it will produce results other than those desired. At this stage of your development it will be of great help to constantly check and re-balance the ego. The ego is the foremost pitfall on the Gameboard, because seeing yourself as separate in any way will instantly divide you from your power. We suggest instead that you focus your energy on your connection to all that is around you.*

THE LADDER OF HUMAN ADVANCEMENT

It is the ego that fosters most destructive forms of competition. Competition for energy is actually a way of manipulating the primary fear to motivate oneself. This form of competition also reflects a core belief in the concept of lack. Some confuse energy competition with a desire to better oneself. Marking your progress against another or yourself is appropriate and an accurate use of energy. It is your judgment around this measurement that one is better than the other that is a misdirection of your energy. It is humorous to us the way you extend this competition into the spiritual aspects. The way you reference vibration indicates your judgment that a higher vibration is more desirable than a lower one. We tell you that this is not the case on the Gameboard of Free Choice. The Keeper recently used an analogy that we agree with entirely. It asks you to view life as a two-sided ladder. Many steps on the ladder are full. Some people are climbing and some are descending. Some are knocking others off in an attempt to reach the top. Some are not even on the ladder but are content holding its feet firmly in place from a position on the ground. It is not until you stand back and look at the ladder from a broader perspective that you understand that each one has a place on that ladder. Furthermore, each one must be in that place for the entire ladder to be in balance. Once the ladder is balanced, and all are working together, then the entire ladder can move forward as one. These are the mechanics of what you call ascension. Your ladders are the spiritual families now in the process of re-uniting. Bless those below you on your ladder and judge them not, for they play a much greater part than you may imagine.

Ladder of Human Advancement   Phyllis Brooks

RE-TURN TO LIGHT

As you evolve to re-turning to your bodies of light, you are learning to listen to what you call higher aspects of Spirit. As we have stated prior, there are many on the planet now opening these channels to allow the natural flow of information. This process that you now call channeling will become even more commonplace and widely practiced. The new words used to describe this process will no longer carry the slant of mysticism implied by the word "Channeling". This will make a neutral space for others to re-claim their truth through this form of communication. Once again, we re-mind you that even with all this progress, this is but a temporary situation. It is simply preparation for a time when you will communicate with energies much higher in vibration. You are now in lesson to communicate with your own higher selves. When you have regained this communication we will be here, ready and willing to communicate as we do now, yet you will no longer need us to re-mind you of your own magnificence. Answers will come to you from the highest source directly through your own higher self. You will have re-membered the part of God that is within you and you will readily express that power. This will be a time when the re-union of spiritual families will culminate in the grand re-union of the Godhead itself on both sides of the veil.

As you learn to walk with your higher self an entirely new way of life will open before you. It will be a way of life much like our own; one that will not include the energy drain of survival. This will be a time when you will walk the Earth because you **choose** to do so, and not because you are attempting to be or to gain anything. This will be the return to the garden that you have set into motion. Such a grand time to be on Earth you have never before experienced. This promise has been given to you for eons, and we now repeat it here. We repeat this promise to you because it is now upon you to open the door to this new world and purposely step in. Many of you see this as a time to come. Some await the brethren from distant planets to start the evolution. Some await the Earth changes as a signal. Some await the return of the Christed one. There are those who have interpreted the information as a day of reckoning and judgment. Others see it as a time of transition from the third-dimension into the fifth dimension. Of these, we tell you that **all** are true. But now we also tell you something more important than anything we have told you prior. We tell you that this way of life is available to you **today**. The ladder is in motion. What you have termed the ascension, has in fact, begun.

ATTRACTING ALL YOUR HEART'S DESIRES

The Earth has just reached a level of vibration that has opened many doors. New healing modalities and other opportunities are surfacing in many areas at this time. Many of you are feeling the pull to make changes in direction. Change is honored and is always beneficial. Judge not your progress with an outdated measuring device. Your co-creative powers to direct your experience at any given moment manifest themselves in your life as change. This full expression of your power allows you the freedom to follow the gentle nudges of spirit, and to co- create a synchronistic lifestyle where all of your hearts' desires are magically attracted to you. Practice this lifestyle and watch as the miracles appear. Look to the Earth for validation of this statement. Change is the way of Mother Earth. She is always in a constant state of flux, and every change results in forward movement. Even those you label as natural disasters, are simply the natural evolutions of the Earth. The Earth is also evolving to a higher level. This process of evolution is not solely for humans, as the connection with the Mother is inseparable. Watch the similarities and honor the pull when it is felt. Because of this new vibration, the Earth will now support these new tools, and your willingness to incorporate change will call them into being. Exciting events and opportunities are about to be placed in your path. It is choices **you** have made that make this possible.

These higher vibrations have also opened the door for each one to begin walking every moment in a conscious spiritual state. In the old vibration this was only attained by a few masters and took many years of discipline. Now, because of your progress in raising your own vibration, this state of constant connection to your higher self is readily available to you. There are many techniques that will be presented to attain this state. The one point we ask you to re-member is to purposely filter all of this information through your own higher self.

BALANCING THE EGO

The ego would have you think that your worth is measured from outside. When you consciously walk as the Spirit you truly are, then judging your worth is no longer necessary. When you view God from your perception, it is generally as a being about which you have no judgment. Who amongst you would judge God? The reason this is so is because you see yourself through eyes that reflect the polarity of the Gameboard. This polarity would have you believe that you are separate from God. We tell you here that there is only one truth, and that each of us carries a special flavor of that truth. We are all parts of the whole that you would call God. See yourself as an extension of the one truth that runs through all things. See yourself as the true part of God that you are. Release the judgments you have about yourselves, for they are a misdirection of energy and make it difficult for you to move fully into your own power. Dare to allow the power within you to show itself.

When that power begins to surface you will feel the flow of Universal energy. It is at these times that we ask you to check your ego. Are you moving into your power while still making room for others to find theirs? Is your message one of personal empowerment, helping all to find their own way Home, or is it one of "follow me for I am now the new leader?" As you re-turn to your power, you will maintain balance only when allowing space for all flavors of the truth to come forward. It is an unfortunate fact that we on this side have had to disconnect some otherwise good messengers because of unbalanced egos. The interesting part is that sometimes the ego is so strong that these messengers have not always known they have been disconnected. You are rapidly approaching a time on Earth when everyone on the planet will be regaining their individual powers of creation. Learning to live in harmony and power will pave the way for a very bright future on planet Earth.

We ask you to re-member that we are here to provide

information to all those requesting it. Please be well advised that we offer **only** information. We do not ask you to follow us in any way, nor do we wish you to even take this information freely simply because it comes from this source. We ask you to run everything through your own discernment, for this is at the base of your true power. Exercise making life choices without the judgment you have used in the past. When something comes into your field, choose it or choose it not, but do not label it as good or bad. These labels reflect judgments that cause a misdirection of energy. Take it as your own, or release it in love as simply not being a flavor of the truth that resonates with you at this moment.

USING THE PRINCIPLES OF LIGHT IN DAILY LIFE

The information that we offer you is nothing more than words on paper. If the words resonate within your own heart, then take them as your own and ground the energy in your life by using the information. However, one does not spread Light by reading words, but by living the ideals expressed by those words. Take the information that resonates with you and find ways of using it in your daily life. This is the part of the Game that can only be played while behind the veil on the Gameboard. This is what you asked to do when you decided to play the Game. Take your ideals of God, and express them in your daily life through your own actions. You see these actions as small and insignificant. We tell you here, there are many eyes on this side positioned to watch this unfold. We tell you that these are the actions that only **you** can accomplish, from the exact position in which you have so carefully placed yourself on the Gameboard. By agreeing to forget your power you have placed yourself in a position to endure many difficulties and you are loved beyond your understanding for this. Let the fears of the past now turn into your greatest strength as you shift your perception and walk fully into your power.

We come to you in many forms, and this family is available to all that request contact. Voice your intent aloud and set it into motion. Release expectations as to how the information will come, and instead, ask that it be for your highest good as filtered through your higher self. To release all expectations is to become the wide-eyed child that you once were. View everything as possible, for this is when we can most readily feed you the information you request.

Have you not noticed that the same information is brought through many channels of differing identities? This is necessary to accommodate your diversity of vibration. Find the teachers that resonate most fully with you. Your need to identify us in terms that fit your ideals severely limits your understanding of the energy. We ask you to use discernment and follow the energy of the message, rather than the words used to describe the messengers. Be advised that we on this side of the veil are much more varied in what you would call personality than are you. Therefore, applying a set of values designed for the human experience may severely restrict the flow of information. There is only one constant that bridges the gap between Heaven and Earth. This is the constant of love. As we have stated prior, this is the base energy that precedes all other forms of energy. This is a universal constant and for you it is a connection to Home. Look for the love content in all communication regardless of its origin. If you cannot feel the love, then release it without judgment, for that teacher was simply not your flavor at that moment.

Because of the higher vibration there now are new tools available to help you realize your own powers of co-creation. It is our greatest pleasure to help you fulfill all of your desires. As you do this, you will begin to create your version of Heaven on Earth. As each one in turn finds their joy, they make it easier for all within their field to find their own joy as well. As soon as each one in turn creates their version of Home, a critical mass is reached and the Gameboard of Free Choice then becomes Heaven on Earth. This is *now in motion. We are deeply grateful to be playing a small part in this evolution.*

CHANGES IN YOUR BIOLOGY

*With these new vibrations your biology is experiencing "density drops" as it begins returning to its former ethereal state. Be advised that this movement will trigger your egos into action in an attempt to ensure your survival. During this advancement your egos may become acutely sensitive. We ask you to be patient with yourself and others as the new paradigm settles in. During the advancement from physical form to ethereal state, you will find most of the resistance coming from within yourselves. Keep a watchful eye on those who see themselves as separate. Understand that it is simply an effect of polarity and help them, if possible, to see unity in all things. Be cautious of those attempting to take your power by promising to show you the way Home. The only way Home is within you, and it is accomplished **only** by finding your **own** power. Let understanding, forgiveness and nurturing replace judgment as these situations arise. Re-member that the ego would have you believe that you are separate and will work to divide you from your connection. Balancing the ego therefore requires a concentrated focus on the connection with your inner spirit. You are now making wonderful progress in this area.*

There is an inner being within each of you that moves with every step you take. Becoming conscious means honoring that spirit. At every moment, realize the connection to spirit and utilize those natural powers. With practice you will soon discover that these are your own powers returning to you once again. Ask that inner being to connect and move consciously with you once more. Make ceremony to celebrate this re-union. Ceremony holds a key to this re-membering, for it invites the biology to play an active role in the process. Your ceremony of marriage is designed to allow your biology to participate in honoring a stated intent to walk in unison with a human partner. Do the same for your intent to re-connect with your higher self.

TOOLS FOR LIVING IN THE HIGHER VIBRATIONS

Much like your human marriages, the real work begins in earnest after the ceremony. Walking together and supporting each other are actions that must be learned and practiced. This is also true of learning to walk side by side with spirit as your partner. This takes conscious effort on your part until it becomes natural. We give you the following ceremony as a tool:

26 SECONDS TO CHANGE YOUR LIFE

At the beginning of every day, sit in silence with your thumbs touching your two middle fingers. The moment you do this you should feel a tingling on the back of your head or neck as the electrical connections complete the circuit. Call on the part of God that is within you and allow a moment to strengthen this connection. Next, allow only thoughts of the greatest of possibilities that await you during the day ahead. See them unfold with spirit always at your side as your partner. See the doors as spirit magically opens them in front of you. Let your mind play with the possibilities. Have fun and return to the state of vivid dreaming you were given as a child. Everything is possible in this state. If there were no restrictions and you were given any wish, what would you create for yourself this day? Do so now. Do this for a minimum of 26 seconds. If at any time you find negative or fearful thoughts entering your field, release your fingers, take three deep breaths and start again. During your day, whenever you are presented with challenges or adversity and feel your energy unbalanced, use this tool to enter this state once again. Just connecting these fingers will instantly call spirit into action to balance your energy. What you will find is that, with practice, you will quickly develop a powerful tool to use in any circumstance.

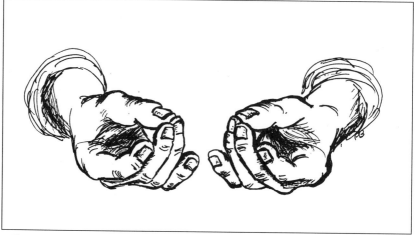

26 *seconds to change your life*   *Phyllis Brooks*

As we have stated before, the connection between your physical and emotional bodies is inseparable. Change in either of these bodies naturally reverberates in the other. During this time of transition there is much shifting in the physical bodies that causes emotional turbulence. Know that this is a temporary condition and lean on those in your spiritual family for support. All in this vast clan are experiencing similar trauma on some level. There are many healers here now to help you move past this stage of discomfort. We have addressed this before and given valid techniques for grounding that will help to ease the discomfort associated with these shifts. We speak of this now because our reference to your history makes this easier to comprehend. Your physical body is changing at the level of your DNA. This is an awakening process that in actuality is a return to the ethereal bodies of which we have spoken. Over time, your need to ingest food products of the Earth will diminish. We ask you to be patient and to not rush this process as there is much to be done on many levels. *During this time it will be helpful to eat with consciousness. We offer the following as a tool for re-membering to eat with consciousness:*

Eating with New Consciousness

Take a moment before ingesting your food and consciously send it energy. Do this in the same manner you would when sending healing energy to another. By sending energy to your food in this way you are actually raising its vibration. You will also be preparing your biology to incorporate your food for the highest good of your Spirit.

Many of you have been using food for the fulfillment of the emotional body. This not only fails to produce the desired result, it is also a misdirection of energy. The following ceremony will set your intent that food be used only for the benefit of your physical body:

Releasing your Emotional Attachments to Food

Prior to eating, take a moment and consciously express your intention that the food you are about to ingest be used only for the nourishment of your physical body. Touch your heart with the fingertips of your dominant hand, then make a sweeping motion across your food. As you make this motion, hold the thought that this food be used only for your highest use in the physical. This simple but powerful ceremony re-minds you to set your intent that the food be used only by the physical side of your body.

Anxiety about Time

Anxiety is a side effect of the higher vibration. It is your new perception of alternate time standards that causes this reaction.

More will be released about this phenomenon in another session to come. All vibrational movement is done in increments that allow time for a response from the Gameboard. View this as you would waves of your ocean. Every wave that comes into the shore is accompanied by a recession of the water. From a limited perspective it would seem that the waves make no real progress. From a higher perspective, however, we see that each wave removes some of the sand on the beach, and every wave makes some advancement toward the land. In much the same way, it may appear that your own progress thus far has been very slight. We ask you to judge yourself not if you feel you are taking steps backward. This judgment carries its own focus and energy and can impede your progress more than you understand. Know the natural order of the universe and allow it to play out for you in spirit time rather than your time. Because of the advancement of humanity it is now easier to see that your progress is substantial. Trust in the process for it will lead you Home.

As you progress you will notice that the tools we give you are valid only inasmuch as they prompt you into your own power. You are more powerful than you can imagine, even in your present state. On a very deep level you know this, yet re-membering these powers in daily existence is often a challenge. It is this use of ceremony that can help the biology to re-member its connection to Spirit. Although we have given you ceremonies that can be of great help to you, we tell you this: These are only to remind you of your magnificence and creative abilities. They are tools to help you concentrate your powers. Change them to suit yourselves if you feel that pull, for it is not our intent to take your power in any way. Use these tools as a starting point until you begin to grasp your own power. We are here to help you re-member and strengthen your own power. The most powerful ceremony anyone can use is that which they set up for themselves.

Your focus is what you are creating in every moment. This simple fact is the basis of the power you are re-membering. We encourage you to center on your desires in life as these will surely find their way to you. Focus is everything. You are more powerful beings than you imagine. Your focus is always the blueprint for your future. Why is it that you try so hard to have different attributes than your parents only to find that you have become them? It is a matter of focus. Why do you always create your greatest fears? It is a matter of focus. Tune your focal point to the spirit within and watch the changes unfold. Walk as though this internal spirit is always on the outside. Greet others as if it were spirit greeting them, seeing only the highest potential in all situations. Venture to locate the spirit within each one who enters your field. Speak to them as if you are honoring the spirit within them in the same manner as we speak to each one of you. Re-member that there is a spirit part of yourself beside you at all times. Your shadow is a re-minder of this reality. Others can only see that part of you if you intend it, for you are on the Gameboard of Free Choice. This spirit moves everything within your field to meet your expectations. Focus your expectations to a higher level and you will then create a higher reality. Your power will return if you learn to walk every step engaging the spirit part of yourself. This is the mastership that is readily available to all who now choose it. Conscious living is now an available reality. It is your next step.

Keep the proper perspective, and above all please re-member that this is a Game. The intent was to have fun. If you are not **enJOYing** the Game then the energy is out of alignment. Align the energy by adjusting your focus on the spirit within. Thoughts are the ethereal energies that create your reality. If you are not happy with your reality then we invite you to think again. Change your focal point and you will change your reality. Do so until the joy in life returns. It is this joy that marks your path Home and leads us all to the creation of Heaven on Earth.

It is with great honor for you that we ask you to treat each other with respect, nurture one another and play well together ... the Group

How do you get to heaven?

Practice, Practice, Practice.

This channel re-minded me of one of my favorite sayings: "The definition of insanity is doing the same thing over and over expecting different results." As the Group said, "If you don't like your reality, then think again."

I know that although I get to carry these messages and make them available to others, they are always addressed directly to me. I have to say that in my life the ego issues surfacing seem amplified right now. Of course it's always easier to see things in others than it is to see them in oneself. It's certainly less painful. Often, the things we are working on diligently in ourselves are the things that most easily agitate us when we see them in others - especially if they aren't working on their stuff! A perfect example is reformed smokers who become intolerant of smoking. With a new paradigm being defined for the ego, this could be a sensitive time for all of us. Years ago, when the Group first came to me, I asked them what I could do to best carry the message. Their first caution was for me to watch my ego, check it, and balance it often. They said that this, more than anything else, has ruined some perfectly good channels in the past. The balance comes from re-connecting to our higher selves and walking in spirit while in biology. So how do we do this?

Practice, Practice, Practice.

Chapter 7

Co-creation

The art of manifesting with spirit

Co-Creation

So often we find ourselves looking for the reason we are here, and what we came to do. We often think it has to do with a specific contract, whereby we will somehow find our illusory golden path and change the world forever. The Group says this is not the case. What we have done, is to script roles for ourselves, setting up appropriate opportunities to walk through Karma. To them, there is no judgment about whether or not we are on our paths or in our contracts. They honor us for simply being. They say that it is in this "being" that we most effectively express ourselves as the pieces of God that we are. They honor us every time we make a decision. It makes no difference if we see that decision as "right" or "wrong." It is the very act of determining our paths that makes us the expression of God that we are. Every decision that we make is our expression of the God within, at that moment. The truth is that we do not have to be anything. We need only to be. When we allow ourselves to simply "be," we begin to re-member Home.

A PICTURE OF GOD

I love to ask the Group about how we got here. They lovingly oblige me, and do the best they can to explain the secrets of the universe to this novice. They say it is beyond our comprehension in our present form to understand simple concepts like infinity, simply because we live in a finite world. Quite often when I ask the Group a question, the answer comes back that I should accept that I cannot understand everything in my present form. This really frustrates me. They humor me a lot and still attempt to answer my questions. After all, they do love to laugh. They also are quick to point out that we are evolving at a tremendous rate and soon will be able to comprehend much more than we now do.

Such is the case with our origins, and the concept we have of Spirit or God. Here, the Group has given me the visual image of a large circle. This circle has no beginning and no end. It is as close as we can come to understanding infinity. This is the expression and nature of Spirit.

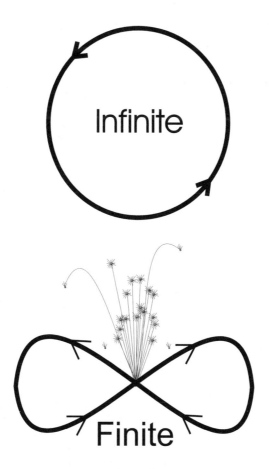

The expression of infinite energy.

Infinite energy crossing itself creates finite sparks. Once the sparks fizzle out, they fall back to rejoin the original circle. For a short time the sparks are finite expressions of the infinite whole. We are those sparks.

The energy of this circle is in motion, yet this motion cannot be defined because there is no stationary part of the circle. In order to accomplish definition it must first become finite.

In this quest to become finite, the circle of God twists in on itself and crosses over to form a figure eight. This figure eight now represents the infinity sign in motion. As the other side reveals itself it can be seen to be in contrast to the first. For example, let's say that one section of the figure eight is white, the other black. This is the beginning of what we call polarity. The original motion of the circle is now traveling in a figure eight and for the first time there is a point where the energy crosses itself. This is the first point of contrast that makes the energy motion definable.

At this middle juncture of the figure eight, energy runs into itself causing sparks that magically fly off into space. The sparks put out brilliant displays of light that last only moments before they burn out and once again recycle into the energy of the circle. These sparks have a beginning and an end; therefore, they are finite pieces of the infinite circle. As such, they can be defined. If you now take your understanding of God or Spirit as the infinite circle, with us as the finite pieces or sparks that flew off, you now have a clearer view of what the Group shows us as our origin.

From this illustration you can see that polarity was a necessary component of this game. In fact, it was polarity that first made it possible for the circle to become finite and thus for the game to begin at all.

These small pieces of the large circle are each of us. Because we are separated from the larger circle we do not easily re-member our origins or our heritage. It is also interesting to note that if you examine any one of these sparks you will see that each one has two distinct sides; One black, the other white. They are exact replicas of the original circle. It could even be said that they were made in its image. Each one of these particles is also polarized with the same magnetic properties carried by the larger circle at the time the sparks were formed. This relates to what we call astrology today. Subtle magnetic fields in the Universe at the time of our birth are carried within our biology as definable attributes. The Group says that we have all the attributes of the original larger circle as well as all of its powers, but since our separation we have not re-membered how to access them.

One of the ways we can access these powers is through Co-Creation. Co-Creation is the art of deliberate manifestation with the help of Spirit. In a nutshell, it is designing every moment of our lives and creating our own destinies with the help of our higher spiritual selves. It is being brave enough to walk purposefully into our own futures, the way we want them to be. It is "being" in its finest form. Most of us carry the seeds of self-doubt and unworthiness, and therefore make this difficult. But the Group says God wants us to have everything our hearts desire. After all, it is our heart's desire that will help us create Heaven on Earth. The Group is very big on Co-Creation, and has given us a lot of information about this subject; how it works, why it works, and why it sometimes does not. They wish to help us "be" our finest expression of God. They wish to help us shine very brightly during our expression as the beautiful sparks of Light that we are.

The Group:

It is with a great honoring that we once again address this powerful gathering of masters. You are loved beyond your understanding for the work that you now do. It is our joy to be here with this information at this pivotal time in your evolution.

Your vision of yourself is tainted with the effects of polarity; therefore, the experiences you carry from childhood have become deeply embedded within your being. This is not always in your best interest and presents problems as you begin to rediscover your powers of creation.   As you become aware that these wonderful gifts await you, your ego steps in to tell you that you are not worthy of such gifts and powers. This is very humorous to us, as we clearly see you for the masters that you are. To view you as unworthy of any gift is simply not within our grasp. If your vision of God were to appear in front of you this moment and ask you for something, would you not give it to Her without wondering if He was worthy of it? We now tell you that God stands before you many times each day, asking you for things that you judge Her to be unworthy of. This is not in your best interest, and holds you back from your manifestations more than you are aware. These are your desires, and God is the one looking back at you from your own mirror.   It is your nature, your heritage, and your right to claim the gifts that lie before you. We are here to help you re-member those powers.

MANIFESTATION. . . HARNESSING YOUR POWER

At the very basis of your powers is the art of manifestation. From our perspective, we term this the art of creation. Since it is not yet within your belief system to allow yourself that power, we will call it manifestation through Co-Creation. There were times long ago when you studied this in earnest, and there were schools that you designed to help in this study. The scrolls containing this information have recently been uncovered, yet crucial parts are being withheld for the purposes of control. Because of this act the information is now being made available to you through other sources that cannot be so easily manipulated. You are the ones who called it back into your field. Through your intent to raise your own vibration you have created the space for this information. As a result of this space a vacuum was formed that pulled this information to you at this time. Because the flow of information was impeded in one direction it simply found other avenues to balance the vacuum. The truth cannot be controlled or manipulated for very long. Nature always finds balance.

Now these same schools have different names and modalities than before, yet the message is the same. In ancient times these schools were a formal training ground and were required for all choosing to walk in mastership. In more recent times this same information was taught in schools where the overall information was broken into two parts. One is the mechanics, (the knowledge, or left brain) and the other the emotional aspects, (or the right brain). It was the approach of these schools to separate the two aspects and then teach integration. This was a direct approach to working with polarity. In polarity consciousness, one sees with eyes enhancing separation. This was an opportunity to present this information to the conscious, allowing the success of the information to bring about the integration. This experiment was successful and very appropriate for the overall vibration of that age. It is still a valid approach and many will yet find it useful. As you move away from polarity consciousness we will offer this in a more direct form. At the time these schools were developed we placed much information into the collective consciousness about the art of manifestation. Humanity as you know it is now taking a very large step into the next stage of evolution. In preparation for these steps, it is now time to once again remember this information. We are pleased to make this available to those who seek it.

WHAT IS CO-CREATION?

The nature of Co-Creation is thus: Co-Creation is when man intends a manifestation and asks Spirit to carry it out. When Spirit intends a manifestation and asks man to carry it out, this you have called Co-Incidence.

It is when Spirit is in the process of carrying out these

manifestations that we align energies in your dimension. You often see this as a hand reaching in and changing your world. You have labeled this with many titles such as "predestined," "lucky," or "unlucky," depending upon your judgment of it. Sometimes you use words like "chance" and "random," and the one that has so much humor for us is when you call it "God's will," for that it surely is. What you see as "lucky" is so often the realization of a manifestation that you yourselves put into motion. These are the synchronicities that we ask you to make room for in your lives. It is very often the manner in which Spirit returns your Co-Creation.

Your Co-Creations are returned in exactly the same vibrations as they are sent out, and this often causes you to not recognize them when they return. You often label these as random accidents. A course of study now coming forward is the art of allowing space for synchronicity in your lives. Through intention it is possible to create the vacuum needed to bring in a synchronistic lifestyle in which nothing is needed from the outside. Such was the life of the master you call Jesus. This is an honored study, and will complete the triad of the tools of manifestation.

Within each of you is a connection that defines you as an individual. This is your connection with your higher self. Although at times you may feel it to be a weak connection it exists always within each one of you. This connection actually defines where one person begins and the other one leaves off. Without this connection the energy would all run together. It is very difficult to use words to describe this connection as each one feels it differently. It is best described as an inner knowing. There are times in your life when you have full connection here. These are experiences or re-membrances, which you know in your heart as the truth. To have another inform you that this was only a dream, or your imagination, would be useless, because this truth is ingrained within your soul. This is a connection you can experience more often than not.

The way to increase this connection is through intention and practice. This connection is the power within. Exercise it, and learn to trust it. It will always connect with your biology in definable areas. We have given exercises to help locate and utilize these areas. Find these areas of your biology and watch for the feelings there. They are accurate barometers, and are the physical measuring tools of discernment. Become well-practiced in this area, for this tool can be used in many ways. We ask you to use this tool of discernment on your own Co-Creations. Spend your energy only on those that resonate within your own heart.

There are many levels of manifestation. The Co-Creation process is the level where you place yourself in the natural flow of the Universal energy and allow your manifestations to be drawn to you. The Co in the Co-Creation is the joint effort of man and spirit. Since this form of manifestation involves spirit, it is a form of manifestation that can only be used for the highest good of all concerned. Man has Free Choice in his manifestations, yet Spirit has a predisposition of light. Therefore, all Co-Creations must be for the highest good, or they simply will not manifest.

THE PROCESS OF MANIFESTATION

The art of Co-creation or manifestation is a very simple process. All it entails is placing yourselves in the flow of energy. As this applies to the human condition, it will be more understandable to you if we break this simple process into basic "rules." There are many of these rules that apply to the art of Co-Creation. Here, we will cover the basics, and discuss things that are often blocks to your success in this area. The illustration of the large circle that the Keeper gave you is valid. We shared this information with him as a visual so that he would readily share it with others. You are all sparks of this grand experiment, as are we on a different level. You have the powers of the whole available to you at all times.

*Re-membering the means to access these powers is the reason we are here at this time. We say to you that this is not about being a success or a failure but simply about **being**. It is the act of deliberate manifestation itself that allows you to be a full expression of God, in whatever form that takes. It is the moment you stand and create your future that is honored. The fulfillment of your manifestations can never measure up to the magic in the moment that you create it. This is the reason that you often find emptiness in your belongings. It is the moment of creation that is your greatest expression of your power. All you need to do to win the Game is to "be." May we help you to understand that you need not "be" anything. To simply "be" is enough.*

ACCEPTANCE

No Co-Creation can manifest unless it is accepted. It is your judgments about yourselves that most often block your manifestations. When you allow yourself to "be" there are no judgments about what you are being at any moment, except for those you place upon yourselves. All judgments are easily traceable back to judgments that you have about yourself. Identify these base judgments and the energy will begin to shift. This is the acceptance, or emotional, portion of Co-Creation. These judgments, or belief systems, are the single most important reason you have been unable to fully re-member your powers up to this point. Changing these belief systems within you is the strongest expression of Spirit within. It is the Co-Creation of the masters.

When you find yourself in judgment about a situation or another, take a moment and look within for the match to that energy. As you uncover it you will find it is a judgment that is embedded within you about yourself. Examine these, and if they no longer serve your highest good then releasing them is in order. Release them in a ceremony of your own design. Trust yourself to design the perfect method, for there are many on your shoulder every moment. Focus on these negative feelings intently. Bring them into view and then trace their energy back to their origin. When you are ready, offer them up to spirit for release. Ask that they be taken from you and release them with gratitude.

Many in this family are challenged by what you have termed abundance problems. We tell you that most of you have no difficulty in manifesting abundance. For most in this situation the challenge comes in accepting that abundance. Many are the times when you send out energy into the Universe in the form of your gifts or your work. When that energy returns in the form of monetary abundance it is shunned or turned away. If you turn energy away it will simply find another path to travel. The effect is that you stay in your lack of abundance while others seemingly get handed things they did not deserve.

The natural flow of Universal energy is about energy finding balance. If you do not accept the energy as it returns to you it will find another path. You look to your neighbor and see them as lucky. Often you say that the rich get richer and that money attracts money. We tell you that an attitude of acceptance attracts the reflection of energy that you call money.

You are the ones that formed the Gameboard and decided the manner in which the Game was to be played. What you call money is your representation of energy on the Gameboard. See that your judgments about money, as a representation of energy, are actually the judgments you have about yourself being reflected in this area. Create the vacuum in the Universe by practicing the art of graceful acceptance in allowing your manifestations to find you. The purpose of the game is to create Home on your side of the veil. Would this not be easier to accomplish with the acceptance of abundance?

All is attainable from within. This is the work that is highly honored. As you free your belief system to accept more of your own power it magically comes to you. Create the vacuum and the Universe will fill it. This is a predestined energy match, and is at the heart of what we term Co-Creation. Therefore, the next step would be to learn to create the vacuum, then walk away with the full knowledge that it is being filled.

TIME LAG

The Gameboard of Free Choice, as you have designed it, has in place a safeguard that has ensured your survival on the planet. This was a necessary component that was added when you shifted to the dense bodies you now inhabit. This safeguard is known simply as the Time Lag.

In the fifth-dimension there is no time lag between the time you have a thought and the time it manifests in your world. This is representative in all of the higher dimensions, because linear time only exists in the lower levels. The challenge comes in the third-dimension, because here you are not yet masters of your thoughts. If instantaneous manifestation were present it would not take you long to manifest your greatest fears and end your lives prematurely. It was necessary to introduce the Time Lag as a way to effectively interface with the third- dimension. On the Gameboard at present there is an interval between the time a manifestation is set into motion, and the time it is realized. It is as if the Universe is asking "are you sure?" The Co-Creation travels through this time lag, and presents itself only if you are the same vibration as when you set it into motion. If at any time you change your vibration or the co-creation during this time lag it becomes negated, and the process must begin all over again. With every moment of this time lag, the universe is making sure this is the correct vibrational match for you. It is during this time lag that so many Co-Creations fall away. For most on the planet at this time, this is as it should be, for it is an effective way of protecting you from yourselves.

The length of the time lag varies greatly depending upon several variables. Those Co-Creations set into motion with strong emotion will move through the time lag much faster. If you set a Co-creation into the time lag with passion, it will manifest quickly. This is also the reason that you always create your greatest fears, for fear is a powerful emotion. Other variables include the time needed for Spirit to position things or events. Although we, as Spirit, do not experience time in a linear fashion, when we intermingle on the Gameboard we do so in the time frame of the Gameboard. Another determining factor in the time lag is your own belief system. If you have an inherent belief that a manifestation will take time, or if you are not ready to receive, it will be delayed accordingly.

There is one other part about the time lag we wish you to know. As you raise your vibrations and the cumulative vibrations of Earth all time lags are being reduced.

When looking at the tool of Co-Creation, it is helpful to remember that the universe is limited to returning only exact vibrational matches to your requests. There is no room for universal interpretation. You will receive exactly what you ask for. We caution you to closely examine your words, because you will receive exactly what you ask for. Choose each word that comes from your mouth and become aware that these words go into your own ears. This is where you start the process of Co-Creation.

We share with you now the most valuable information about Co-Creation. The Universe has only one answer to all of your requests. All co-creations are answered with the phrase:

"And so it is. . . . . . . "

GETTING THROUGH THE TIME LAG

When getting through the time lag there are basically three components. They are **Think, Speak** and **Act**. The basis of

Co-Creation is held within these three words. We shall walk through them together.

THINK: BECOMING MASTERS OF YOUR THOUGHTS

Thoughts are the beginning of each Co-Creation. We ask you to be conscious of every thought that you allow space in your head. Until such time that you master your thought processes, we ask that you control what you allow into your thoughts through your other senses. Consciously choose all thoughts entering your head by becoming aware of what you allow to enter through your ears and your eyes. Every person or thought you come into contact with will share your vibration momentarily and shift your own vibration ever so slightly. Be conscious of the vibrations you allow into your field. Choose carefully who you spend time with, and what you allow into your head through your senses. Choose what you read and what you watch on television. If it empowers you then take it as your own. If you are not comfortable with the thoughts that enter your head then turn your eyes elsewhere. The direction of your focus is entirely your choice in every moment. Be aware that there are many who would ask you to Co-Create their manifestations by appealing to your senses. One method is what you call advertising. This is an area that is well informed in the mechanics of Co-Creation. This is not to be feared or judged. At all times you are in control of your own Co-Creations simply by being aware of the process. Tracing the energy to its origin will always reveal true motivation. Make your choices, fully informed of motivation, and this will provide balance. Also, be aware that all ideas you accept as your truth are the beginning of the Co-Creation process. It is not possible at your stage of development to fully control all thoughts entering your head. You are, however, in full control of the thoughts that stay there. Choose these well for they contain your next reality.

SPEAK

As you designed the Gameboard it became necessary for you to ground yourselves to the Earth in order to complete your task. As this happened, it became necessary for you to find other means of communication. The languages you developed are very confining and did not make space for "words" that spirit uses. Therefore, it is wise to study the words leaving your mouth at all times. Listen to your words as they relate to the phrase: And so it is...

The Universe is only a mirror of your energy and all is returned exactly as sent out through your intent. The actual words are not heard on this side of the veil, yet the energy is accurately reflected in your choice of words. This is the energy of the answer the Universe always gives to each request it hears. Therefore, it is very appropriate to choose carefully the words that are ex-pressed, as they relate to the answer.

For instance, it is common practice to ask for things to be given to you. We re-mind you here that to ask for anything is actually a statement of lack to the Universe. What the Universe hears you say is that you are not whole until you have something added unto you. Do you really want the Universe to grant this request? This exact scenario is played out more than you may be aware. This is what we speak of when we say that you are much more powerful than you imagine. When you say you want this or that the Universe responds with: and so it is. . . and so you will "want" it for a very long time. Study the energy of your statements prior to setting them into motion.

We offer you here a re-minder of the most powerful words that can be uttered for the purpose of Co-Creation:

*These are words of **Gratitude.***

To give thanks for something moves the process of manifestation from a state of doing to a state of being.

This is only a slight shift of perception, yet it is the most powerful of tools that we can give you to help you to re-member. To give thanks for anything is a statement to the Universe that it is **already** in your possession. The Universe naturally responds with: And so it is.

ACT

There is action required on your part to bring into effect all manifestations. In some instances, there is actual action necessary for you to do in your dimension. Spirit will line up the energy for your manifestation, yet it is yours to provide all the action required on your side of the veil. It is only when you have done all that you are able to do that Spirit will take over. On all Co-Creations there is a point when all the necessary components are in place then and it is time to give it to spirit. This in itself is a very strong action, and requires the trust and faith to let go, and let God. This action is sometimes one that is very difficult for you. So often we see you standing so intently by the door, waiting for it to open. What you do not understand is that the door opens toward you. In your anticipation you are often standing in its way. Have the strength and courage to let go. When all is done, and everything is in place, let Spirit have it. Release it to the universe with the full knowledge that it will return complete. This is applying the final act that sets all manifestations fully into motion. It is the act of faith. Walk away knowing that it is already yours. This knowing is the beginning of the vacuum that pulls the rest of the manifestation into place.

Co-creating with your Higher Self

Another important part of this fine art is the understanding that it is not possible to Co-Create for another. Co-Creation is accomplished through the connection with the higher self. Since it is not yet possible to maintain a connection with anyone else's higher self, it is likewise impossible to Co-Create for anyone other than yourself. It is possible to Co-Create in unison, as long as each one is Co-Creating for themselves. It is possible to create space for another to accomplish their visions, or to receive healing. Yet in all instances it must be their individual choosing to walk into that space. The words "self first" have a predisposed meaning, yet if you can remove the judgment it reflects the natural flow of energy within the Universe. All things must first come from within. To Co-Create for another is to put another before oneself. This is a misdirection of Universal energy and will not produce the desired results. There will be much more on this in future communications. Have the courage and self-worth to place yourself first in all matters. Only by centering your own energy will you find it possible to give of yourself fully.

There is much more information given that is available to you now. We ask that if this resonates with you, seek it out and create the vacuum that will bring it into your field. Many of you will be teaching this to others, for it is time for such information to be presented on the planet.

You have traveled such a very long way to get to this moment in your history. What awaits you is beyond your comprehension. Co-Creation is but a tool that opens the doors that will lead you home. The gifts that await you are many, and we are here to help you find your way. Practice discernment at every opportunity, and find the messages that are yours. Fear not, for fear creates its own energy and can divert your focus. Know that you are well protected and guided every step of the way. Listen to that guidance within and find your connection to your higher self. You have earned the right to be here at a very exciting time in the history of the Gameboard of Free Choice. You are on the brink of winning the game.

It is our deep love for you that inspires this re-minder. We ask you to treat each other with respect, nurture one another, and play well together... the Group.

I have to tell you that this was an interesting channel for me. The Group has given me so much information on Co-Creation already that it was very interesting to see the parts they picked out as the most important at this time.

When I sit to write these messages there are many Group members that come to play. The opening paragraphs of this message are mine alone. Yet when I got past that point in writing I became aware that a member of the Group that I call the scientist was with me. Those of you who know me can easily tell the energy difference. This information came through me as though he/she were sitting on my shoulder whispering in my ear as I wrote. I call these "gentle nudges." This is the actual process that we do during the private "re-membering" sessions. They are with me, nudging me here and there, guiding the conversation but not talking directly. This happens to all of us at times, but we often call it our imagination. This is humorous to those on the other side.

I thought I'd share this because they say that more people are beginning to channel now. As we share information we can make it easier for each other. From the Groups' perspective the art of listening to spirit is not mystical at all. They say we can bring in a lot of valuable information, and also have a lot of fun if we don't take ourselves too seriously.

Sometimes I really think I am here just for their amusement. Not that it matters... I really love the sound of that laughter.

Chapter 8

Synchronicity

Allowing room for spirit in our bubbles of biology

Steve and Barbara in Amsterdam, Holland

Synchronicity

L ightworkers everywhere are now raising their vibrations at an increasing rate. Things are moving faster now than ever before, and change is becoming the norm. I have often wondered about how this takes place. If we become aware of the mechanics of this event, can we make the shift smoother? The Group responds with a resounding "YES."

The Group:

We are truly honored to be here with this gathering. There are many now being drawn to this vibration that will play key roles in the shifting of planet Earth. You are honored for feeling the pull and following it. Let us be the first to welcome you Home again. This is a time of positioning and Re-membering the cast of characters. A third wave of planetary healers is now being positioned. This was not foreseen. This positioning is taking place with the highest of honors, for you have moved yourself into this reality quite unexpectedly. Your role in this Game was to be a finite expression of the infinite spirit. It is with great universal excitement that we see you now re-membering your powers of the infinite and pulling this into your own finite reality. This added twist to the Gameboard of Free Choice has brought a new reflection to the outcome of the Grand Game of Hide and Seek. This has implications far beyond your understanding. We truly honor your work behind the veil. It could have been accomplished no other way. Through your eyes, you view yourself in lesson, struggling to move forward. It is now for you to know that we see you having already won the Game. The colors you will carry for this accomplishment are great and will be with you always.

The Earth is also now positioning Herself, and although She may remain quiet for a time, there is still much work to do shifting the energy within. *The Gameboard, as originally designed, was due for a dramatic shift that would have discontinued biological life on the planet. Through your work in raising your own vibration this has now changed. The work you are in the process of doing at this time has to do with softening the transition to the higher vibratory levels. As you are aware, you are fully capable of affecting matter in your own reality, including those changes yet to come.*

So often you look outside of yourself for answers. We tell you once again that all is within. If you wish to change the world it can be accomplished within, one heart at a time. When you link your hearts and hold hands in intent your powers become amplified exponentially. This was the truth that prompted these meditations as monthly messages. Due to the rise in your individual vibrations the expected energy shift can now be greatly softened. This channel of information began as a way to soften these changes within the Earth. Your Co-Creation was to heal the Earth. As this Co-Creation was set in motion through your intent, you allowed Spirit the room to bring it into your world in its most effective form. Therefore, the manifestation of this Co-Creation was to heal yourself, for as we told you, your connection with Mother Earth is inseparable. As you remove the blocks within your own emotional bodies you then become able to carry more healing energy into the Earth. There are many that would view this as a change of direction or focus. We tell you it is simply leaving room for spirit to work in your life. When you ask a question, you often define the manner in which the answer must come, this limits your highest return. In this time of great change, those who are capable of releasing attachment to the outcome will feel most at ease. This is a key to living in joy in the coming times. Become comfortable with change and your world will be seen through the eyes of a master.

Change is now upon you at a quickening pace. If you wish to equate this to something in your reality look to your computer technology. It is an accurate mirror at this time. This change has started with the outer bodies of your being and is now working its way to the inner levels. During the past six years you, as a race of humans, have been reworking your outer, ethereal bodies. There have been many changes made to these parts of yourself that you have been unable to see as of yet. When the time comes you will be very joyous of these changes, for they will ensure your survival. Since these bodies are on the outer portion of your being, far from the physical, they have not caused much discomfort. Instead, these changes have shown in your reality as an intensifying of the spiritual self. They have manifested as a global spiritual awakening and a quickening of the pace. Another dimension of time is now revealing itself to you as a direct result of the changes to your ethereal bodies. Those of you at the leading edge of this vibrational change have become acutely aware of the other realities that surround you. For the first time you understand that there are parts of yourselves that extend into several other dimensions simultaneously. The next step in your evolution will be learning to control movement in these dimensions. More will come on this later.

THE CUTTING EDGE OF CHANGE

The changes being made to your bodies are now moving into the emotional and physical levels. As they move into these layers, nearer your core essence, you often feel discomfort. The emotional and physical bodies of your existence are very closely linked. Because of this arrangement it is only possible to change **both** concurrently. Unfortunately, this amplifies the discomfort associated with them. This is the greatest challenge for those of you in the higher vibration, and those you call Lightworkers. Because the Earth is still in the lower vibratory levels, those leading the path to higher vibrations are experiencing what they often perceive as difficulties in their lives. The higher perspective is that these souls are clearing a *path for many to make these transitions with greater comfort in the future. This is the real work of the Lightworker. You are blazing a path of light with every difficulty you encounter. Know that the work you are doing within, though it may seem trivial to the larger picture, is actually setting the stage for what has been termed the ascension. You are literally changing the world within your own hearts.*

On the physical level your biology has begun shifting at an increased rate. The seeds for the new biology are now firmly planted and are beginning to germinate. There have been some important biological changes taking place at this time. Energy has been released from a higher aspect into your field and the change has now begun in earnest. Within these changes lies potential for emotional discomfort, as your emotional foundation shifts along with the physical. Many Lightworkers felt particularly vulnerable during this shift as their biology has been reacting to this change.

You humans do not usually honor yourself with a slow, gradual assimilation process. Instead of getting on the gradual escalator to higher vibrational states, you usually jump on the express elevator, transcending hundreds of floors at a time. And then you expect to exit the elevator and walk normally. When the elevator stops and the doors open, you find it difficult to function in this new environment. During these times you are very sensitive to energy. It is quite common here to unknowingly pick up the energy of others, thinking it is your own. Compound this sensitivity with the increasing vibrational rate of the planet, and you have great potential for imbalance. This imbalance often finds expression through heightened emotions. It is common to experience emotional swings. Please understand that these are a direct result of your eagerness to move forward quickly and nothing more.

INCREASED SENSITIVITY TO ENERGY

As your biology continues to alter you will find yourself becoming more susceptible to the subtle magnetic fields caused by these planetary alignments. We ask you to take this in your stride, understanding that it is caused by the physical limits of your biology. This is a time to lean on each other for your emotional balance. You have asked to move forward and this is now manifesting on a global level. Mankind is moving into its next evolutionary stage, and through your expressed intent you are leading the way. From an historic viewpoint this will happen in the blink of an eye. It is already in motion within you.

Those who are tightly attached to their ideals may find themselves struggling against the flow when, in fact, they themselves actually asked for the flow to change direction in precisely this manner. Those who are comfortable with shifting realities will make the best of each moment and will seemingly glide through life effortlessly. Now is the time to open to this way of life. It is a way of life with room for change, and room for the hand of God. It is a lifestyle that truly reflects an honoring of spirit.

There are those of you at this time that state intention to move in a forward direction and then release it to spirit as a part of the Co-Creation process. Spirit then begins positioning to allow for the most efficient way for this to manifest for you. As it begins to form, you often resist its placement, for it represents change and the unfamiliar for you. This is counter-productive to your original Co-Creation and is often the source of pain. At times this even negates your Co-Creations.

From this side of the veil it appears that you do not trust spirit to manifest your Co-Creation. There is no judgment in this, yet it effectively limits your efforts. You ask to be led onto your path, and when the doors open to that path you then ask why your life is falling apart. *Spirit must be invited into your lives before interaction is allowed from this side of the veil. This is the prime directive of Free Choice. Many times spirit begins moving things to line up with your choices, only to hear you express your fears at the changes in progress. This has the same effect as revoking your permission for spirit to do the work. Then you express impatience at the lengthy process. Change is not something that humans do well; yet it is the only way to bring you the things you ask for. Relax in the process and understand that all change will only be for the highest good. In all cases, change will lead to something better.*

These are the times we wish to just hold you and love you and let you know how important every little move you make is to the greater good. There will come a time when you will be able to see all of these things. Then, and only then, will you see the grand scale of the work you are now doing. You have agreed to participate in this grand experiment for the good of all that is, and our honor and love for you cannot be placed into words of expression. It is only because you have asked that we give you this information at all.

DEFINING YOUR ENERGY FIELD

The first attribute to center within is clearly defining your own energy field. To know where you begin, and another leaves off, will allow you to release all that does not belong to you. Intertwining your energy with those around you was easier in the lower vibrations and became a way of life for many. You have often done this as an act of sacrifice, yet you also found a certain comfort in not having to define your own energy. In the higher vibrations you will find true strength comes only from within your own field. If you allow your energy to be invested in others, your focus will scatter and there will be little of yourself to give to anyone. Those that clearly define themselves will be the ones that others seek out for guidance. This is the higher purpose of what you have called the ego. The ego is a tool for personal energy balance. Much like a balloon, it is possible to inflate or deflate the ego, and in so doing a balance of energy can be maintained. Allow the ego to over-inflate and it will overshadow your entire being, allowing you to see only the ego. Likewise, if there is a lack of ego, you often invest in others' energy fields to compensate. There are those among you now teaching this balance. These are the ones who move forward in their own power while allowing space for others to find theirs. These are the next wave of teachers and healers. These new teachers and healers are the ones who walk firmly in their truth while allowing space for their students to find their truth within, even if it may seem to be contradictory to their own. They will stand out as a higher vibration and all will be attracted to their fields.

Once you clearly define your field it is easier to release your attachments and just allow yourself to be. It is often your attachment to the outcome that hinders your forward motion. Likewise, when you intricately define the manner in which spirit must manifest your Co-Creations you are revealing your attachment to the outcome. Herein lies the dichotomy that sidetracks many of your Co-Creations and severely limits the way we can place things in your field. We ask you to re-member that the greatest act is the Act of Faith. Once your Co-Creations are set into motion, release them fully and keep the posture of trusting Spirit. This attitude holds the key to the most productive Co-Creations, for it leaves room for sprit to work in your life. With practice this brings about a synchronistic lifestyle where everything seems to magically drop into place at just the right moment. This is leaving space for spirit in your life and leads to a joyous existence.

If you look back on the events that have shaped your life thus far, you will see that at first you judged many of these to be setbacks. We say to you now that if you had not judged these to be setbacks and set your posture thus, you could have incorporated this positioning more rapidly. As events unfold in your life your natural first action is to judge them as good or as bad. This is an effect of polarity on the Gameboard and is a natural human reaction. This judgment allows you to define your feelings and posture on this event. We tell you that it most often begins a circle that is difficult to escape. Here is the typical scenario: An event enters your field through your eyes and ears. It is judged as a setback. In an attempt to understand how this could have happened you begin to verbalize it again and again. Here the cycle is set into motion. As you speak it, the vibrations of your words reflect your judgment of it as a setback. As these words leave your mouth they travel around your biology and find their way into your own ears. As this vibration enters, it becomes amplified and is sent out to the universe as a Co-Creation. As we have spoken before, the universe has only one answer to any request put to it. That answer is: "And so it is." What began simply as a misperception is now a reality because you have just Co-Created it.

A Posture of Gratitude

We ask you first to consider applying the immediate posture of gratitude to all events as they enter your field and reserve judgment for another time as to "Good" or "Bad." See yourself as playing the Game. Feel the excitement and anticipation as it becomes your turn to draw the card and make your move. If the card you drew appears to be a setback, allow your first thought to be one of gratitude that it is your turn to play. When you intend to move into your joy, do you watch for outside influences bringing that joy? Perhaps it might be the path of least resistance to find the joy in the things already within your field. It is this slight adjustment of perception that holds the key to higher vibration. Know that there are no "wrong" moves on this Gameboard. Please re-member that it is a Game to be enjoyed. You are protected beyond your understanding and it would be very difficult to really hurt yourself unless it is your direct choice. Play the Game with passion and anticipation, watching for the hidden gifts in every square you encounter, even if that square presents itself as a setback.

One of the Keeper's favored sayings is: "Pain often accompanies growth, but misery is optional." It is polarity on the Gameboard that filters these events with an expression of "Good" or "Bad." Do not let this fool you. Each event that enters your field contains "The Gift." Each of these events has within it the seeds of growth that will take you to your highest expression of the Spirit. Find "The Gift" and you find your true power. Find your power, and you are in your joy. Find your joy, and you have won the Game, for then you become a full expression of Spirit. Learn the rules of this Game well. This information is available from many teachers in many forms at this time. There are new teachers and methods in your plane now to show you the rules of this grand Game in a light that you will accept. Seek them out and use your inner discernment to find the ones that most easily resonate within you. The master healers among you will be the first to seek out these healers and teachers, for they know how to use them as a reflection without the need to give their power away.

THE PATH OF LEAST RESISTANCE

At this time in your history it is helpful for you to move forward with a minimum of resistance. Here we ask you to adjust your thinking to locate your path of least resistance. Prior to each phase of the Game you placed many crystals in your path. These crystals carry the vibrations of Home and resonate with you so deeply that it will be difficult for you to rationalize them away. As you find these crystals they will bring you joy and you will note the pull of your heart-strings. Follow these pulls and they will lead you to the center of your contract. They are re-minders you have placed here for *yourself. Make space in your daily life for spirit to speak to you through synchronicity. Your path of least resistance will allow you to move freely upon the Gameboard at will. This is your true expression of the one spirit within all of us.*

You see yourself as being in biology with a core essence. We envision you as a grand master spirit playing inside a temporary bubble of biology. If you misplace that vision call upon us, we are available to you at any moment simply by asking. If you dare, take a moment and look at yourself through our eyes. We are here only to help you re-member your magnificence.

It is with great love for you that we ask you to treat each other with respect, nurture one another and play well together . . . the Group.

Synchronicity Exercise at The Lightworker Spiritual Re-Union, Vienna, Austria

Barbara and I were on a flight to Tucson when I began to write this message. I only got a couple of pages written on the short flight as I kept getting sidetracked somehow. It was as if the Group wanted this to filter into my life over the next few hours in a way I would re-member.

We got to Tucson and rented our car for the weekend. We had

a two-hour drive ahead of us as we were doing a seminar in Scottsdale the next morning, before returning to Tucson for another on Sunday. We were cutting it a little tight, but we would still get at least one good night's sleep if all went well. Upon our arrival in Scottsdale it was midnight and all we wanted was to sleep, but there was a snag. The night clerk informed us they had just given away our room, even though it was guaranteed and confirmed. He said they had another, smaller room for us, which he assured us would meet our needs. He also informed us that it was a smoking room, but not to worry because they kept their smoking rooms well disinfected. My first reaction to that was anger, but just as I was about to share my feelings with this smiling gentleman, the Group tapped me on the shoulder to re-mind me of what we had just written on the plane about synchronicity. "Okay, so maybe this is just a game," I thought to myself, "I guess I'll play this out and see what happens, you know the whole bit, to allow room for spirit to work through synchronicities and all that. In fact I'll even go all the way and bless the situation." I muttered under my breath that I was only doing this because I was tired and it was late. I was definitely thinking like a human, eh?

Recovered smokers are usually the most discriminating, and Barbara and I are no exception. So we found ourselves in the elevator holding our breaths (literally) in anticipation of what was to come. We both smiled at each other as the floors went by, but we were both really attempting to gauge what the other's reaction was going to be. The hallway to the room smelled and irritated our senses a little, but it did not prepare us for what was to come. When the door opened to our room we were met with a stench that was overwhelming. It was obvious to me that a poker party had just left it moments before our arrival, and that housekeeping had immediately rushed in with four quarts of "disinfectant" to make our stay more pleasurable.

Ahhhhh yes, the acid test - and I was really tired. "Okay, Group, is this some kind of joke?" (Laughter from over my shoulder somewhere.) I could feel the words of discontent forming in the elevator on the way back down to the lobby. Barbara was silent - no smiles this time, and I didn't have to guess what she was thinking. The gentleman behind the counter magically anticipated our lack of excitement and quickly handed me another set of keys to try. This room was better, but we were still unhappy. We decided that it was late, so we would just have to tough it out and then check out in the morning before the seminar to allow time to air out.

In the elevator I had a quick flash of sanity. Re-membering the writing, I was able to step out of myself for a moment. I now saw that I had a choice as to how I wanted to play this Game. I could get angry and voice my discontent, or I could just flow and play the Game, focusing my energy on the important parts. The moment I made the decision to flow, there was a release within me. I no longer had any attachment to the outcome. As we passed the front desk to get our bags the night clerk said that he had a surprise for us. Since we were the ones renting the conference room the next day he was going to take care of us in style. He had arranged for a suite of rooms for us at another lodging five minutes down the road, compliments of the hotel. I cracked a smile and the Group winked.

The story does not end there.

The next morning, I got back to the hotel an hour prior to the seminar to check out the conference room. After announcing myself the day clerk forced a grin that told me she was hiding something. They did not have us down for a conference room. I began to feel the familiar words of discontent forming in my mouth, yet I held them inside. This time I wanted the manager. After waiting 20 minutes for him to locate our forever-lost paperwork, I finally took him aside and told him that we were expecting people to arrive soon. "Where would he like us to put them?" I was envisioning a Sword of Truth seminar in the lobby, but quickly realized that was out.

At that moment he smiled, and then led me to the perfect room. Although Barbara and I had to set up the room ourselves, it was excellent accommodation. The very moment we got things into place people began to show up. The small co-incidences that continued throughout the day were too numerous to relate, but the Group had a lot of laughter. Again, the charge for the room was less than half of what I had originally negotiated (there is that laughter again).

This was a prime example of using the information the Group had just given. I was very much aware that if at any time I had allowed myself to turn sour this whole situation could have gone another way. I could have very easily been standing at the front entrance of that hotel giving people refunds as they entered. Instead, I just played the game.

Thoughts create reality. If you don't like your reality... think again.

Chapter 9

The Illusion of Fear and Evil

Or, be careful where you point that thing!

The Illusion of Fear and Evil

A TRIP TO THE TAVERN OF LIGHT

Some time ago when I was first awakening to the information from the Group, I had an experience that I would now like to share. During a meditation the Group took me to my special place. This resembles a tavern of sorts. This is really funny because even when I drank I never frequented bars or taverns. However, this place is very special indeed. There are wooden tables and chairs, and it feels very homey. They take me there often. It's a "Light Bar" where the beverage served is always "Light". They laugh as they explain that this was one of the reasons we humans are so attracted to the word Light in our three-dimensional world. They say our higher selves have been planting seeds with these words such as: Lite Beer, Lite potato chips, Lite this and Light that. We were purposely raising the consciousness of the planet by dropping seeds with the words we were using. Then, to top it off somehow, our children even ended up with Lights in their tennis shoes. As it became more fashionable to use the word Lite/Light, we were actually planting seeds to prepare ourselves for the important messages to come. Now that mass awakening is taking place, hearing and reading such words as "Lightbody" has become more acceptable. This is a direct result of the conditioning our higher selves set into motion. It's no wonder they took me to a place I know as the Tavern of Light.

Every time they take me to this Light Tavern I know I am in for a treat, as they often reveal to me the secrets of the universe. On one of these visits they showed me many things that helped to explain some of the situations I was wrestling with at that moment. I remember thinking to myself "It's so simple." I made a mental note to keep in mind how simple the universe really is. If it's getting complicated then we are getting off the track. This turned out to be a great rule of thumb for use in my life following that experience.

The Group have taken me to this special place many times and they are nudging me now to tell you more about one of these visits. The strange part about my visits to the Tavern of Light is that they don't allow me to return with the direct knowledge of what they shared with me. Most of the time I become aware that the memory of my experiences has been erased. It seems I am to return only with the feelings that are associated with these events, as well as the seeds themselves. This doesn't bother me because I know they are planting seeds, and in fertile ground those seeds will surely sprout. The task they now give me is to find that fertile soil in others and pass these seeds along. It is my contract, and my joy, to ground the energy in this fashion.

There was one trip to this place that particularly stands out, because it was the first time I became aware that I could interact. Before, it felt as if I was more of a spectator. The Group took me on this wonderful journey, and though we spanned many continents we never left the tavern. They shared with me awesome secrets of life. I vowed to return with at least the feeling of how uncomplicated it all is. Then I became aware that it was time to go back. We were saying our parting goodbyes in the street just in front of the tavern when an idea struck me. As I was walking away I turned back to the Group and said: "I know you're not going to let me re-member any of this, so would you give me something I can take back to share with others?" In that instant they handed me what I can only describe as a woodcarving. It was about three feet long, with the following words carved upon it:

"THE OPPOSITE OF LOVE IS NOT HATE, IT IS FEAR."

Just then, I heard a barrage of voices, all speaking this same truth in different words. It was overwhelming, but I distinctly re-member picking out one familiar voice; it was Roosevelt saying: "The Only thing to Fear is Fear itself." I began to see how this same message has been given to many other people. Now it was being given to me. The only difference this time, was that they had wrapped this truth in a different "flavor."

My very first reaction to this was to question the awkward use of the English language. After all, the words given to Roosevelt were so much more elegant. The Group immediately roared with laughter because, of course, they "heard" what I was thinking. The humor was, they said, because *I* was the member of this group who was responsible for the proper use of the English language. That was my job and it would be helpful if I got to work so they did not have these problems. I got the message loud and clear and joined in the laughter.

This was my first experience of what I have come to regard as "negotiating" with the Group. I thought it was really cool that I could bargain with them like that. Actually, they encourage my requests now because we have moved to a different level of communication that's much closer. I'll give you one guess as to which one of us moved.

Through the tools they have given, and their encouragement, I am finally beginning to take my place as a member of this special, but anonymous, Group. I described this same experience in previous writings, and used it to illustrate what they referred to as the Dark Side. The Group has planted many seeds since then, and they now wish to revisit this misunderstood enigma of what we call "evil."

1945 THE BEGINNING OF PLAN B

Before I bring in the Group they want me to address one other point. In an earlier message, I was writing a message from the Group when they casually mentioned that this big shift of humanity began 53 years ago. I did not have any clue as to what they were talking about, but simply wrote it as it came through. Upon editing, I wanted to alter the figure 53 to read, "about 50 years ago." Since they are always on my shoulder, the Group immediately stepped in and said, "No, leave it as is." When this has happened in the past, they've explained that they sometimes use codes and specific vibrations to send messages on many levels that I don't understand. I thought this was another of those occasions and let it go.

About a week later, I heard from three people in different parts of the world. One person was from Austria, one from Africa, and the other from the state of Washington in the USA. All acknowledged the figure of 53 years and said it was an important date for them and thanked me profusely for including it. (Thanks, Group) They all said the same thing. They pointed out (at the time I wrote this) that it was 53 years ago that Adolf Hitler left the planet. I thought that was interesting, but still failed to see any real significance until I asked the Group. To my surprise they said that Adolf Hitler's departure from the planet represented an enormous milestone for humanity. As a result, that moment would live forever because it marked the last time that humanity would give away its individual power on such a mass scale. This date has great spiritual significance because it marks the start of our awakening process. Now, when new leaders come forward and attempt to assume the posture of supreme power, that old vibration is no longer supported on this planet. Although these people will still come forward they will not last as we are no longer willing to give our power away in that manner. The time had finally come for each human to take control of their own environment, instead of looking to any leader to do it for them. There was great joy from those watching this Game at that moment. From a higher perspective, that event in 1945 was the first time we humans made a conscious decision to carry our *own* power. This opened the possibility for us to divert our paths from "Plan A" to "Plan B."

Hitler played a pivotal role for humanity by pulling the pendulum so far to one side that it might never have to move in that direction again. Moreover, the six million Jews who perished as a result of the holocaust gave mankind a truly wonderful gift. They had all contracted to give their all so that humanity could awaken and forever alter its course. Let us never forget that - or them.

The Group always speaks about the honor they hold for all of us as we make the decisions that define our own worlds. This, they say, is the greatest expression of the God within each of us. When we make statements of power, such as we did in 1945, we are beginning to create Home on this side of the veil. What does the Group say about this? Just two words: "Welcome Home".

The Group:

There is reverence and joy within each of us as we address this assemblage of Lightworkers. It is your own intent to move forward to the next level of existence that has allowed us the opportunity to present this information. You have moved to a new vibration and this is applauded throughout the universe as it opens doors for all to redefine new paradigms for themselves. As each one of you redefines your role you naturally move closer to the grand re-union of what you would call the Godhead. As this takes place you will begin to feel the true vibration of Home on planet Earth. As each one purposely adjusts the tone of their own heart they will find true harmony, and then they will truly create Heaven on Earth, one heart at a time.

THE FAVORED STORY OF ALL TIME TO COME

What we will now discuss is the second part of a two-part message. We gave the first installment some time ago, but it was important for those seeds to take root before we completed this information. This is now accomplished. As the Keeper said, it was this action of taking your power that set plan B into motion. *This was the deciding factor. Even though it would be years before the Gameboard would be measured, this was **the** pivotal point in your history. There are many among you who are the Keepers of the Story. You are the Librarians of this grand library called planet Earth. This is honored work, and there will be much for you to accomplish in the near future. For now, suffice it to say that the events put into motion in 1945 will be a favored story of all time.*

At this time among the masses there is still a backlash of energy that makes you wish to look the other way at these events. There are even those who have claimed that these events never happened, but were fabricated for the gain of a few. We tell you not to worry, as these times on the Gameboard are permanently inscribed in the hall of records, never to be lost. We also tell you that there will soon come a day when humans can access these records and follow the story of an awakening of empowerment. Many will use the story of the Game of Free Choice to illustrate emerging power.

This is the reason we, as the family of nine, are here at this time. We are here to re-mind you of your own power. We are here to help you awaken to the mastership within each of you, and thus enable your next step in evolution. This will be the first time on the planet that you will be able to walk in the bubble of biology fully intact with your whole being, including the part you call your higher self. The changes are now taking place within your own biology to prepare for this. These are the times when you are taking your power and holding it as your own, rather than looking on the outside for something to be added unto you. Understand your wholeness.

SOFTENING THE EARTH CHANGES

The Earth, too, has been making vibrational shifts in preparation for this time. These shifts show in your field as Earth changes. These adjustments will continue at an ever quickening pace, yet there are two forces at work that are greatly softening their detrimental effects on humanity. One is that you have now begun to take your power and use it to define your own world. Time and again we have told you that you create with every thought. You have begun the process of becoming masters of your thoughts and, as such, have begun using these tools to ease this shift. Your full reality is but a compilation of the thoughts you have allowed to reside within your being. As you become aware and begin choosing which thoughts you allow within, you make choices that clearly define your next reality. You are now becoming more aware of this process and how you can direct your world with it.

We have given a morning exercise that many of you now use to clearly set the tone of your overall day. This is a morning exercise of holding your thumb to your two middle fingers [middle and ring] on both hands while allowing only thoughts of the highest proportions for the day to come. Do this for 26 seconds and it will effortlessly direct the course of your entire day. As you now understand, these exercises work, for they allow you to experience your own power of creation. You have only to re-member to use them. As you use your own powers, and purposely create your own reality, you ease the tension that has been building within the Earth. This is the reason we ask you to go within and do the real work of Lightworkers. It is this clearing of your own field that results in positive changes for the Earth. The connection to the Earth is much deeper than you imagine. Clearing the energy blocks within you directly benefits the planet as a whole. This work is only possible in biology, and is the nature of the Game.

The second main force at work is that humanity as a whole is becoming more comfortable with change. Change is difficult. Because of polarity on the Gameboard, there is a natural feeling of resistance in response to all change. Still, as you stretch yourself to accept the challenges of change, you see that you are larger than you imagined.   You wonderful humans are more resilient than you believe. We tell you that all change ultimately has a positive result. As we have stated before, the universe is about balance. When there is a pull in the fabric of the universe, it naturally responds and fills the vacuum. When there is a push, the universe creates a vacuum to make space for it. This is reflected in many ways within your own daily lives. If you look back only a few years you will see that the blending of humanity has made great strides.   It is this blending

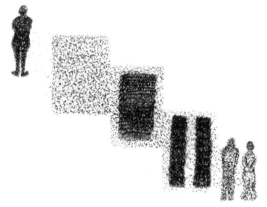

Brenda Pakkala

across the illusory lines of race, ideals and so-called origin that has paved the way for the next step in the evolution of humanity. This was the decisive factor in the defeat of the one called Hitler, for those ideals were contrary to the natural flow of the universe. Humankind awoke to that truth and took back its own power. Even to this day there are governments that strive to preserve the purity of certain groups of people. These are all destined to fail because they oppose the natural flow of energy-balancing within the Universe. The flow of Universal Energy is always seeking balance, and any attempts toward segregation are in opposition to that natural flow.

THE COLLECTIVE CHOICE

During the time of Hitler you made choices that were in accord with the flow of Universal energy. Had your collective choices gone the other way, it would not have been possible for the contact and truth you are about to receive. The contact you are about to receive will help you in ways you cannot imagine at this time. This you have earned. Had your ideals been abrading the fabric of the Universe for this entire time, it would have changed the contact that you are about to experience with the enlightened ones soon coming to your Gameboard. You have created this great time by your actions on many levels.

Now is the time when we ask you to step forward and fully accept your own empowerment. Understand that you are in charge of every moment of your life, and that the events you see around you are only reflections of what you have allowed in your thoughts. Direct those thoughts as you would the pieces on a Gameboard. We ask you to begin to hold your power on any level. This is the reason we ask you to use discernment in all areas of your life, and to be well practiced in the art of listening for the answers within. Even the messages from us we ask you to filter through your own heart, and take only what resonates within, for this is the form of listening that will connect you with your own power.

In your work as healers and teachers we encourage you to act appropriately, for it is no longer appropriate to take the power of another, even though they may hand it to you freely. Help those who come to you find their own power, and nurture them as they become comfortable with their own wings. Utilize your own healers and teachers for validation and help in re-membering what you already know. Seek guidance and validation rather than answers. As others enter your field, choose only those that leave you more empowered by their reflection. This is the mark of mastership. Those who truly understand the connection with all life know that to empower others is to empower the whole. A fully empowered whole is what you would call Heaven on Earth, or Home on your side of the veil.

THE FLIP SIDE OF LIGHT

So often this power is misdirected. The laws that govern this power do not dictate the direction in which the power is focused. This is the Gameboard of Free Choice, thus you have Free Choice as to where and how you use your power. We have spoken before of the use of this power. Many of you on the Gameboard view the lack of light and then give it life and power by calling it darkness. This darkness does not even deserve the word for it is not something to be described. It is the **absence** of something. It is an illusion created by polarity on the Gameboard, but it is a necessary component. Without contrast it would not be possible to define that which you call light. The interesting part to us, is that you are **so** powerful, that you easily create this thing that you call evil. We tell you that it is only a reflection of your own fear that makes it so. This is truly a testament to your powers of creation.

Observe the sea as it rushes to shore with great force. As that force dissipates it then flows backwards into the ocean to find a natural balance once again, for this is the way of the universe. You would not look to this process of ebb and flow as good and evil, yet this is what many have done. This is a natural process, and a necessary component of your three-dimensional world. They are simply parts of the whole. Follow the flow of energy and the universal flow, and discover your own truth about this thing called darkness. Darkness is not the opposite of Light, it is simply the **absence** of Light.

We see most often that it is your own judgments that hold you back from the advancement you seek. We have offered you the tool *of discernment so that it might be easier for you to see how it can be to make choices without judgment. If you point to anything and say that it is bad, you have placed an energy around it that gives it life. Even as you attempt to walk away and separate yourself it follows you because there is a link to your own energy from which it gained its life. These energies then become very real indeed as the power they now hold is equal to your own in all respects. They are in fact offspring of your own thoughts. You always create your fears. This is the mechanics of how those fears manifest.*

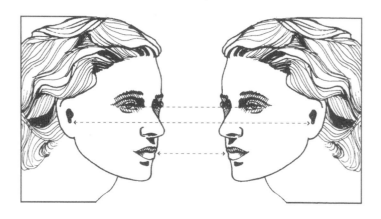

Careful where you point that thing!   Eve Meng

Observe that in your bubble of biology your eyes point in one direction and receive vibrational input from only that direction. Similarly, your ears point in the same direction as your eyes, as do your nose and mouth. All of these are very powerful vibrational sensors. We ask you to keep in mind that this is the way you communicate with the three-dimensional world that we call the Gameboard of Free Choice. It is through these sensors that you not only receive information, but also send vibrations out to the Universe as part of the co-creation process. We ask you to be very selective where you point these sensors, as the direction in which they point is the direction of your next reality.

We re-mind you that in the tools we have given you on the art of Co-Creation we explained how those manifestation, accompanied with emotion travel through the time lag faster than others. Create with passion in your heart and you will never be denied. The one emotion that sets many manifestations into full motion is the powerful emotion of fear. The most misunderstood emotion is often the most powerful. For many on the planet this is the one directive that determines the course of daily life. There is no judgment about this, only observation that it is not returning the goals you have set out to attain. It would be a more natural reflection if one would view the emotion of fear as being similar to the natural backwash of the ocean; it is simply part of the whole. In many ways the emotion of fear has been a necessary and useful tool. In the same way that darkness is only a lack of light, fear is only a lack of information.

Your wonderful human nature has you searching every corner of your existence for answers to all your questions. Your inherent belief is that if you have the answers to all questions then all things will be added unto you. We tell you that all are asked to carry only small parts of the puzzle, and that together we make the whole of creation. Yet each one of you is completely whole within yourself and has access to all other pieces through your connection to one another. This is the reason we are a Group of nine, each one offering their own field of expertise. This, which we offer for your individual discernment, is a blend and balance of information. It resonates within your heart because it is of the one truth that is from Home, with the flavor and pull of your original spiritual family. Lean on those around you to fill the void of information and watch the fear dissipate.

FEAR IS ONLY A LACK OF KNOWLEDGE

We make note here that those with deep understanding and knowledge are very spiritual in nature. They possess the inner *knowledge that lends itself to a connection to all that is. Understanding leads to a connection with all things, and therefore to more knowledge. It is this knowledge that leads to the spiritual connection. It is impossible to keep the two separated. Since real knowledge ultimately leads to spirituality, is it not possible to say that understanding leads to love? It is here that we make the connection to show that Fear is the opposite of Love, for Fear is the lack of understanding, and Love is the abundance of it.*

In learning to use fear as a tool, we ask you to view it as an emotional signal that there is simply missing information. First, feel the fear and learn to see it as an opportunity for advancement. Then step into that opportunity with quiet resolve as it presents itself. Once you have felt the fear and moved on it no longer has power to deter you from your joy. Do this often and soon, instead of feeling the fear in the usual way, you will feel excitement at the opportunities about to unfold. Contrast this feeling with that of love. As we have shown the Keeper, fear is the direct vibrational opposite of the emotion love. When you are in the state of Love, all is available to you and there is no lack of understanding. See this as the ebb and flow of the ocean with both having their rightful place and use.

We truly honor and love you. This love is more than a family connection, even though that is very strong in itself. In your present state we cannot make you understand how honored you are for the path you walk. The veils are thick and often block your true powers. We give you tools, we help you to re-member, and we guide your steps when asked. Ultimately, it is you that must decide which path to walk and even which tools to use.

It is up to you to walk through your fears. We see the many mixed signals you receive, each one claiming to be the right and only answer. We have great compassion for you. What you <u>do not</u> see is that this time is of great importance to the entire universe, as yours is the only Gameboard of its kind. This was an experiment of *truly grand proportions. This fun little Game of God hiding from Herself, searching for Himself, has led us to uncover some very profound truths about all that is. It may help to know that there are times in what you would perceive as your darkest hours that often provide answers that could come in no other way. Understand that you are never alone. Feel us around you and if you wish our guidance we are always there for you. Lean on us and we will always be there to hold you. All you have to do is to ask and to re-member. If you find yourself with an absence of information, ask and we will be there with answers to fill the void of your fear. You are loved far beyond your understanding. You are part of a very wonderful family of Light. We are so very proud of you. The light you now carry will be seen for eons.*

We ask you to treat each other with respect, nurture one another and play well together... the Group.

I wrote this message while on an airplane heading to West Virginia. Barbara does a great job of shielding me from distractions at these times. She instinctively saw what I was doing and, not wishing to interrupt me, pulled out a book and began to read. As I was writing, however, I could feel something building in her. Finally, she couldn't wait any longer and interrupted me to show me something in her book. There, in a book on prosperity, was a chapter on the illusion of this thing that we call evil. We were both amazed at the synchronicity of this, because what she was reading and what I was writing said almost exactly the same thing with just a slightly different flavor. There was one line that I will share with you from that book:

The ancient Hebrew word for evil is AVEN, which literally means "Nothing."

There is that wonderful laughter again. . .

Chapter 10

The Children of Crystal Vibration

Live Channel, Sudbury, Ontario

Steve offering help with the channeling exercise at the
Scepter of Self Love, Sudbury, Ontario, Canada

The Children of Crystal Vibration

*D*uring our two-day seminars we have always included a live channel from the Group. These have proved to provide very intimate connections between the family at these events and the Group. In speaking with people afterward we usually find that many received things that we were unaware of during the channel. It is as if the Group offers many messages using the same words. I am including some of these live channels here because the Group also answers direct questions at the end of these sessions. Some of them are very interesting.

This particular channel was a very special channel for several reasons. The first time I went to Sudbury I was struck by the sheer quantity of rock in this location, and the eerie kind of energy that seemed to emanate from it. Having never before experienced such energy in rocks, I found it difficult to understand why the rock in this area should be so different. It was as if something was happening within the rock itself. Our host at this event was a Master Healer named Claire Gibb. Claire explained to me that the reason there was so much stone here was because Sudbury itself is the world's largest nickel mine. She said that eons ago a meteor had struck this area, leaving a crater that became the town of Sudbury. Somehow, the heat from that meteor had brought all the nickel to the surface.

I asked the Group for an explanation, but got no response. I returned home with very few answers, just an eerie feeling. Returning to Sudbury a few months later, Claire picked us up from the airport. Driving to our hotel, I again experienced that same eerie feeling. It felt as if the stone was shifting density and I was tuning in to that somehow. Once again, I asked the Group for an explanation. This time they responded. I saw the Grand Master of Timing as his finger was being lowered. I knew that sometime during the seminar I would be given the information. It was during the live channel on the second day of the seminar that all was explained.

The Group:

We are honored to be here. This is such a wonderful gathering of Lightworkers. You have no idea how far your beacons shine. You look around you and you see yourselves and each other, and you think that this is all there is. We tell you that what you do here goes far beyond what you can see with your eyes. You are loved throughout the Universe for what you are doing right here. The new paradigms you are creating are great and will extend far beyond your understanding. Take a moment and let the energy settle in the room. There are some here who are only just getting used to our energy. We ask you to breathe it in and relax. It will be here for those who choose it. We honor your choices, for they are the greatest expression of the God within you, as is your own individual discernment. What we offer you this morning is for your own discernment. We ask you once again, take what is yours and leave the rest.

We wish to tell you something about this place for it is very special. Many of you have known this for some time. The creation of this place occurred at a time when the Earth was cooling. It attracted a meteor to help facilitate the plan. As the meteor hit, it did so in such a way as to bring the molten rock to the surface. This was for the purpose of balancing the rotation of the Earth, for as the Earth rotated, it was necessary to put a certain spin on it. The meteor that hit this area achieved this. Through the ages, certain people have been drawn to make their home in this area. By lending their vibration in this fashion, they have helped to stabilize the planet. In many ways it is you that have shifted the energy of Mother Earth.

A BULLS-EYE FOR THE DEATH STAR

Your decision to move as a planet from plan A into plan B has had a significant effect on this area. What we will tell you is that there was a magnetic resonance here. Viewed from outer space, this was seen much like a bulls-eye. For this was the place where Mirva was scheduled to hit. Mirva was the sister meteor to the one that originally formed this crater. Mirva's purpose was to bring about the conclusion of the Gameboard as scripted by the original Plan A. This would have effectively removed all life from the planet. This scenario has played out five times prior in the history of the Gameboard. This time was to be the last and final effort to complete the Grand Game of Hide and Seek on the Gameboard of Free Choice. This location was the exact place this scenario was to begin. And now you have changed that. As a result, the rock in this area is changing density as it is no longer needed to mark the spot in the same fashion. As the energy continues to soften it will attract other Lightworkers. This area has become a beacon of wonderful energy and it is you that have made it so. We love you so. We honor you so. Now, the entire Gameboard and the Universe can evolve to the next level and enter new paradigms.

The magnitude of what you do here goes far beyond what you see. And we love you for it. We see you very much at the tip of a triangle, moving through old paradigms and re-arranging them. You will be as the atoms at the tip of the arrowhead, breaking the energy and changing the paradigm at the very beginning. We tell you this so that you will understand the changes that are taking place within you, since it is these changes that enable you to do this work of changing the world around you. It is as you begin this journey that you feel the most resistance. Once you begin shifting the old energy the world and paradigms around you will begin to change as well. You feel these changes in your lives and you think that it is about you. You look around you and wonder if this is all there is. You wonder why you are different, and yet you still have the courage to continue. What you do not see is that behind the tip of that arrowhead, the entire Universe is also changing, moving to a new level that you are helping to create. You are the forerunners. You are at the very *tip of the triangle that is rearranging this energy. And we tell you this: The colors you have already earned and carry for this decision are far brighter than you can possibly understand.*

Many of you are just beginning to see that you are masters working behind a veil that keeps the truth of your mastership from you. Even though you are starting to get glimpses of this, still your ego will not let you fully accept this truth. In many respects this is as it should be, for the balance must be maintained. But let us encourage you to feel your mastership, because it is both your heritage and true power. It is who you are.

Many of you are feeling the nudge to move from one location to another, one job to another, to redefine your life, some in subtle ways and some in drastic ways. We ask you to move slowly when honoring these nudges that emanate from your higher self.

The new energy on the Gameboard is beginning to take hold. This means that many choices will now be available to you. We ask you to finely tune your art of discernment about which choices to make, for this is your greatest tool. New information will now become available to you. As the Earth increases its vibration, the overall collective consciousness is beginning to support new modalities that will be used in healing and teaching.

We thank you for this, for it is you that have made it so. You walk the Earth behind the veils, unaware of who you really are. There is great humor about this on our side as we see you pass each other on the street, looking into each other's eyes, not knowing who you are. We tell you that many times you have done great things together, and that you have planned this in such a way as to allow it to unfold. We cannot help but love you for it. You are so much a part of us. And you are now beginning to let us become a part of you. And for this we are thankful.

There are several who have come here today, and three in

particular, who have asked for healing. You have always looked to us for this healing and we tell you that you are the ones that are most capable of it. We will give you a little exercise to use so that we can show you your own power to some small extent. As you sit in this circle we ask you to begin running energy in a clockwise direction, and to let the energy circulate from one into the other. You do not need to touch each other, just feel the energy move between you. To those of you who have asked for healing, we ask that you breathe this energy in as it comes past you and let it run through your biology, for it is yours in that moment, if you so choose it.

THE SECOND WAVE OF ENERGY

There are many new things that are coming to the planet; one in particular that we wish to help you usher in today. You stand at the threshold of the second wave. We have been giving the Keeper little bits and pieces of information about this. And yet the fullness of what we are calling the second wave of empowerment will not be known for several years, for it is only in hindsight that it will be fully perceivable. Yet we tell you it is happening now.; that it has, in fact, already begun. Some of the markers to look for have to do with finding your individual empowerment. Instead of follow the leader you will now play a Game of follow yourselves. You have no idea how powerful you are even as you are about to reclaim your mastership. You are about to walk once again into your own mastership and claim it as your own. This can now be done safely without disturbing your biology or the Earth. Look to the people around you. See that they are there for a reason. You are here to hold the energy for one another.

Many of you here now are from the time of Atlantis and Lemuria. Because of the trauma associated with these past incarnations there is a seed fear that rises as you get close to this energy. You have a tendency to pull this energy inward and direct it upon yourself, thinking that you have to hold it back. The last time you had full use of this power you capsized the world. But we tell you that the seed fear is there to provide a balance to your own energy. Check your motivations against this fear and then move past it. For as all fears are nothing more than a lack of information, so we ask you to simply supply the information.

OTHERS ON THE SAME CHANNEL

There are others in this room receiving the same information that we give to the Keeper. This is humorous to us because it is almost as if we are hearing an echo of ourselves. We tell you that there are many more in this room who can readily tap into this energy, and we make it available to all who choose it. It is earmarked by the distinct feelings of love. Feel the love as it vibrates within your heart. We are the resonant vibration of this great family of Light and we are grateful for this opportunity to present this information, for it allows us to advance in ways that you are not aware of. For us to play a small part in bringing information to you is the greatest honor we can have. And as it is above, so it is below. Because of your movement on the Gameboard and your own evolution, we also move to higher levels. We are connected, we are one, and we have held the energy on our side of the veil for you.

As you can see in the reflection of your own environment, much is shifting around you. People are reacting differently to you. For what once was feared, is now known. People are now beginning to look past the fear and see the light in your eyes. We ask you to carry the energy, and to make available to others the information we offer to you. For as they see the light behind your eyes they, too, will want that light for themselves. Now is the time for you to plant the seeds of the new energy. Such a grand time to be on the Gameboard!

Many of you have looked at your lives and believed that you are in turmoil. You wonder why you are here. You wonder what your purpose is. And we tell you that many of you have already accomplished what you came here to do. For many of you simply came to see if you could re-member to make the choices you have already made. And we now tell you that because of the new vibration on the Gameboard, and because of the ushering in of the second wave of paradigms, it will not continue to be difficult for you. Feel the energy as it is coming around you now. For as you move through the first wave into the second wave we are able to offer you more comfort. Expect to feel good, for it will be so. The discomfort and the emotional turmoil that you have been through were absolutely necessary, for this was the only way that you could effect what you have done. And we thank you for it. For not all have been as brave as those of you who have chosen to be on the Gameboard and be at the cutting edge of this energy for all of us. You, whom we call Lightworkers, are altering the vibration for all to come. The colors you wear will reflect this always. And we thank you.

THE CHILDREN OF CRYSTAL VIBRATION

Contrary to what the Keeper had thought, we have information that is appropriate to bring forward at this time. We tell you that the rewards of your efforts will be readily seen in the reflections of your own children. The difficult times were necessary, for as you altered the paradigms you also made the Gameboard a safe place for those that will follow. It has been a difficult shift for many, especially on the emotional levels. You have followed your heart and we honor you for that. We tell you that as a result of all your hard work you have now made it safe for those to follow of Crystal Vibration. They are preparing now and will soon begin entering en masse. It is your work that has allowed this to happen. And as you choose to walk into plan B, and as you choose your own ascension process, many of you will be forming your own crystalline bodies and crystalline

structures within your biology. We look at you and we see the turmoil, we see the difficulties that you have to go through in order to
do this. We also see the way in which you are changing your own
world and the manner in which you are expressing that God part
within you; the way in which you are claiming your mastership, and
we realize that you have won the Game. We applaud you for it, for
you are the masters that have agreed to walk and hide your spirit in
these wonderful bubbles of biology.

There will be much information to follow about the crystalline
structures and about those of Crystal Vibration who will be following soon. They carry the new vibration of the planet. They carry the
next important step to that which you have labeled "light body." It
is a returning to the same biology, the same ethereal bodies you originally came in with. You have chosen to play the Game of Free
Choice on the Gameboard of Free Choice in dense bodies of biology, and to hide your mastership from yourself in order to see if God
could find Himself in Herself. You have done well. We ask you to
view yourself as the infinite part, for that is what you really are. You
look at your biology and you see the part of you that is finite. We
tell you that you are in fact the finite expression of the infinite
Creator.

THE VEIL IS THINNING

As the veils are beginning to lift and become transparent you
have the opportunity to consciously connect once again to that part
of you that is all-powerful. And we have given you information and
exercises, and we will continue to do so for as long as you ask, for
it is our greatest honor to do so. It is what you would call our
"higher" purpose. You look around and you see that your life may
be mixed up, or that your relationships are not as you wished them
to be. You may feel the vibrational mismatch of your work and wish
to change everything in your field. We tell you that as long as you

are following your heart you are connected to that part of yourself that is all-powerful. Trust it, allow it to be part of you, and the connections will become stronger. Much as the Keeper talks about exercising the muscles, we are going to ask you to open the conscious channel. Allow yourself to walk every day with spirit, always being there, being conscious that you are connected, and that what you see in front of you is not all there is. And soon what will happen is that you will learn to live your lives from the other consciousness. You will learn to direct your lives from the higher self. This where you are going! And at that point your biology will begin shifting again to accommodate the new energy.

MAKING SAFE SPACE FOR THE CRYSTAL CHILDREN

Those of Crystal Vibration must have a very safe place in which to bring in the seeds of the new humanity. We ask you, brave warriors of the light, to protect these children, covet them, hold them dear, and make this planet safe for them. For this is your contract. In doing this you begin the process of transforming your own biology to the crystalline structure and connecting to the grid which will connect you to all. There will be many questions to come about the crystalline nature and about those of Crystal Vibration, as we have termed it. We will not bring in all of this information, for others will come forward with many more flavors of the truth to balance this important information.

As the Children of Crystal Vibration begin to take their place there may be anxiety associated with what you see before you. It would be common to look at this new breed and think that you are being left behind in some way. We tell you that this is not the case. For this too is available to you in this lifetime if you choose it. Indeed most of you in this room have chosen it already. We look at this and we see the tremendous amount of energy required for this transformation and we thank you. Because of your choices, you have opened the door to allow room for the next vibration of children to enter with the new seeds of humanity.

*You are loved beyond your understanding. You are honored for the Game you play and the way you play it. The energy in this room is awesome. It is a reflection of your energy. You, who have allowed us to be here. And we tell you that what you feel in your heart is not just our love, but a reflection of your own. And as the smile comes upon your face, understand that this is what **you** feel like to us. We embrace each and every one of you. And as you go around hugging each other, know that we are here, embracing you too. We will now spend some time answering some individual questions. There are questions forming in your minds that are beneficial for others to hear. We ask that you share these if you feel the gentle nudge.*

LIVE QUESTIONS TO THE GROUP I

At all seminars, after the live channel, the Group loves to answer direct questions. Here are just a few from this seminar:

SHIFTING OF THE POLES

Question: Are we to understand that it is no longer necessary for the poles to shift on the planet?

Answer: *To clarify. Plan A, as you have written it, was about the poles shifting on the Earth as a result of cataclysmic events. If this had occurred, the pieces on the Gameboard would have been scattered to the winds, so to speak, and then reabsorbed into the whole to find new homes. This location was a landing place for what has been termed the Death Star, the asteroid that was to hit the Earth and help shift it on to its side. This was the resonance attraction and the original reason for the formation of the rocks in this area. The poles no longer will be shifting and that is the reason the rocks are changing density now.*

CRYSTAL CHILDREN VS. INDIGOS

Question: How do the Crystal children differ from the Indigo children?

Answer: The Indigos are much as you are, in that you have been the tip of this great arrowhead changing the paradigm and ushering in the second wave. It was necessary for the Indigo children to come in and completely shake off the old ways with their wonderful obstinance and their preference for re-evaluating everything. They came to plough the field for the planting of important seeds. They are doing a wonderful job, but there is much still to do. The indigos will open the door in such a way that the new paradigms will be possible. Again, it is not for us to reveal all of this information, but we will give you some indications of what this looks like. The Indigo children know who they are. The children of crystal vibration are of a much higher and more subtle, energy. They have not the need or the direction to create new paradigms or shake off old ways, or to re-arrange the structure around them. Theirs is a more peaceful way. These ones know the natural flow of the energy in the Universe and emulate that flow with every action. They are fully connected to their higher selves and are walking with spirit on their shoulder every step of the way. What they are bringing in has to do with seeds of the new evolution. In your current form of biology you will need to undergo a rejuvenation process and DNA reactivation to reach the next level of human evolution. These children will be born with this higher vibratory status. The work the Indigos will be doing will fully pave the way for this to unfold.

RESEARCH ON NEUTRINOS

Question: [At this point I found myself pointing to a section of the room even though the question had not been asked.]

Answer: There is a question forming over here that we wish to answer even though it has yet to be asked. We will address this now because it has to do with the shifting rock formations. Because of the nature of this area, there is some very special research going on here. We ask you not to be in fear of it, for there is no danger. It holds the possibilities of some great discoveries. By looking within the smallest particles your scientists will discover even smaller particles. This will open a door in an area of science and help you understand the much larger particles. The study of the very large and the very small will lead to a basic truth about the Universe you have yet to know. We ask you to fear it not, for the energy of fear can block the actual experiments and taint them. These places have been provided for this study.

[This was in reference to the neutrino study labs. Neutrinos are the smallest particles known to man. The abandoned shafts at the nickel mines in Sudbury have turned out to be the ideal environment in which to study these particles.]

INEXPLICABLE FOOD CRAVINGS

Question: I have noticed that recently I have had cravings for certain foods I would never eat before. Is this because of the biology shifting?

Answer: *As your biology shifts this causes many changes within your physical form. This shift is necessary in order to create a safe environment for the seeds of the crystal children to sprout. These are the seeds of the new biology you are shifting toward. As you move to these lighter ethereal bodies, your body will call for what it needs to make this shift. The cravings you speak of are actually your body balancing itself. We ask you to listen closely to your body, for it knows what it needs to make these shifts. Resisting will only cause unnecessary energy problems and friction. We ask you during this transition to please re-evaluate some of the judgments you may have* *about eating. There are many who think they must eat meat. There are some who think they should not eat meat. We ask you not to think, but to listen. Remove your judgments long enough to let your body balance and it will do so.*

TOOLS TO REJUVENATE THE BODY

Question: Are there tools that you can give us to help us regenerate our bodies?

Answer: *Yes.*

[Here the Group paused.]

We ask the Keeper to relax.

[I found myself agitated at this point and needed a moment to settle down. The Group had been showing me some of these changes on the plane but I had not understood their meaning. I was afraid this might come up and I would be unprepared. When the question was asked it knocked me slightly off center. What they told me in that moment of silence, was to relax and trust them to find the words. I did as they asked and this was their answer.]

The time is not fully right for this to be revealed. However, we will tell you some general directions in which to start looking. Your DNA is not of what you think it to be. The DNA that you see through your microscopes and with your measuring devices has 2 strands that intermingle with each other. You call this the double helix. We tell you that at this moment there are in fact 12 strands already present. The unseen strands are magnetic in nature. This is the reason you cannot perceive them with your measuring devices. Much like antennae, these magnetic strands pick up the vibrations that direct your body how to grow. It is your thought process that sends signals to these antennae. As these signals enter the system they check the current road map of the DNA to see what the new building structure *is going to look like. With each cell that is replaced in your biology a new person is literally being born. The outcome of **who** you become in each moment is governed by the thoughts you allow to reside in your being. In the past, your thought processes have generated all patterns received by these antennae. This is the reason that what you think of yourself dictates who you will become. Because as you think, so you are. Previously, each cell in your body was replenished every seven years. Now this process is speeding up and will continue to advance.*

The Grand Master of timing is holding up his finger at this point indicating that it is not time for full disclosure of this information. But we have given you clues and ways and directions in which to look. It is important that these seeds of information have time to germinate.

THE MIASMA OF CANCER

Question: In this area there are a large number of cancer cases. We have always felt that this is due to the mining in the area. Can you speak about this and tell us if there is anything we can do besides move away?

Answer: *The energy that was here was of a very negative nature. This was necessary to create the target for the plan A scenario. This energy has hung heavily over you for a very long time. Most of you have become accustomed to it, yet still it takes its toll. Additionally, the Nickel mining industry created as a result of the first meteor strike has released many pollutants into the atmosphere. This has created a miasma, which is a physical expression of negative energy. The miasma around the target area actually nourished some who feed off negative energy. Others have been adversely affected through physical ailments. Those that felt repelled but did not leave quickly manifested physical ailments, many in the form of what you call cancer.*

If you listen to your heart and follow the energy, you find your-self in the universal flow and everything comes easy. It is at the base of the information that we taught you about co-creation. This is at the base of a synchronistic lifestyle; of following where you are to go in each moment. And yet, your judgments sometimes keep you from doing that. And you are honored and loved. But if you continually ignore the energy and allow your judgments to take you to a place that does not feel good it will show in your physical body.

MAGNETS AS HEALING DEVICES

Question: *Can you tell us about using magnets as healing devices?*

Answer: *There are some difficulties arising in this field, for magnetism is not fully understood. You have not connected all the dots on the lines at this point and it makes the outcomes of some of your experiments difficult to trace. For instance, you know the effects of what you call gravity. Yet you do not understand its magnetic implications. When you do, much more will be revealed. We therefore ask you to be patient with the mixed results that the scientists will be getting. The way to proceed at this moment is with caution. There is much that can be done, and part of what you will accomplish is a feeling of well-being with magnetics and healing. In some ways you are treating the symptom and not the disease. This usually leads to a dependence on the remedy instead of a cure. Many times the body needs relief from the stress of pain long enough to find its balance. For these and other similar uses in chronic pain relief, magnets have great merit. The orientation of the magnetic field plays a far more important role than is presently understood. Also, you are not yet aware that magnetic fields affect your emotional body.*

FUTURE EARTH CHANGES

Question: Can you tell us if there will be any Earth Changes in the near future?

Answer: *First, we will tell you that no one, including us, can foretell the future. You are changing it constantly and therefore you are more in control of your future than you may realize. There is only one rule on your Gameboard: "In all matters there will be Free Choice." It is your choices that will determine your future path. All we can do is tell you the direction you are pointed at this moment.*

There are changes coming to the Gameboard of Free Choice. The Earth is shifting. You are not the only ones that are changing physical bodies. The Mother is as well. Even though you have softened the energies a balancing must still take place. There have already been Earth changes and much of the process has begun. By advancing your own vibration you are assisting the Mother to shift from her dense body into her ethereal body. We also tell you that many will be leaving as a critical mass of soul-weight on the Gameboard is reached. This is as it should be, for many of these beings will play important roles in the ascension from the other side of the veil. Others will go through a metamorphosis and return to Earth to help birth the new biology as Children of Crystal Vibration.

We honor you for the Game you play. You have done well. We ask you to look into each other's eyes often and re-member your true heritage. For it is in the eyes of each other that you will always find our reflection. Each one of you is of the family of Light. Please know that we can be called by each one of you. The connection you have with us is within your own heart. If you ever lose your way or cannot re-member, please look at yourself through our eyes. We love you so. It is with our greatest pleasure that we ask you to treat each other with respect, nurture one another and play well together.

We are complete... the Group

Chapter II

The Flow of Universal Energy

Live Channel at The United Nations, Vienna, Austria

Universal Energy

O ne day we received an e-mail from a Lightworker in Vienna, Austria, telling us that if we were ever in Europe she would like to host a seminar for us. I wrote back and told her: "Thanks, but at this time we have no plans to go to Europe." About a week later she wrote back asking what it would take for us to come. At the same time, another person wrote asking if I would consider coming to Holland. I could tell that the Group was at work here and they were not going to let this alone until we set it in motion. It was not long before we decided to do a three-week European tour, beginning with Austria, then on to Holland and finishing in Denmark. Our hostess in Vienna was Lourdes Resperger.

During a phone call to arrange this tour, Lourdes casually asked me if we would mind speaking to a group at the United Nations. I almost dropped the phone. (Now I knew the Group was really at work here!) It turned out that she was a member of a group at the United Nations called the VIC Esoteric Society. This Society was founded by an Astrologer in the early 70's and now offers opportunities for people working in different areas of the UN to connect with others of like mind. Occasionally they bring in speakers. From the time Lourdes first e-mailed me she was concerned about organizing such a gathering, as she had never done anything like this before.

There was a time when Barbara and I had accompanied Lee Carroll to the United Nations in New York. It had always been my wildest dream to follow in his footsteps and be asked to speak at the United Nations.   Shortly after the first conversation we had with Lourdes, she informed me that she had just been appointed as the new president of the VIC Esoteric Society. There was that cosmic wink again!

Barbara and I arrived in Vienna after dealing with several

customs officials. It's not easy traveling internationally with an Excalibur sword. After riding on the underground subway, sword and all, we arrived at the Vienna International Center. This is actually an International city within the city of Vienna. There are several buildings that make up the Vienna International Center. This is a home for the United Nations, UNIDO (United Nations Industrial Organization), The International Atomic Energy Agency, UNOV and CTBTO. Upon entering the main lobby area all the international flags could be seen hanging from the ceiling. It was quite a sight. We went up several floors to the room where the channel was to be held.

At 5:30 there was a big influx of people as members of the various organizations ended their workday and came to hear the talk. Many were there from the United Nations, UNIDO and the Atomic Energy Agency. When the introductions were made and I began the talk there were about 70 people in attendance.

I shared with them how this whole thing started for me and how the Group first came to me. I spoke about the messages from the Group and the basis of the work we were doing. I was surprised to find that most of the people in the room had been reading the Beacons of Light Meditations for some time. It seems that the Messages from the Group had been passed around in the internal e-mail system of the UN for several months. They were ready for the information and very anxious to see us. It turned out that we were a validation for them as well.

We were a little anxious about meeting Lightworkers from different countries, but after about five minutes of talking and looking into the smiling, eager eyes in the audience, I realized that we were Home and this was family. I shared the Sword, Scepter and Quill with everyone and explained how we use each tool in our seminars to illustrate the information that the Group has given to us. It was very special to see some of these warriors of the Light holding the Sword.

I had been very concerned about jet lag and having to present something so important to me only a few hours after getting off a plane. To my surprise, the moment I stepped up in front of the crowd the Group came in and I felt the familiar energy filling my being. Jet lag cannot compete with that.

After a short break I began the channel from the Group. They had decided not to let me in on what we were going to talk about. Normally I would have been very nervous about not knowing the subject of the channel, but on this occasion I was simply too busy to worry about it.

The timing of our trip put us in Vienna during the Kosovo crisis, which was only a few hundred miles from our location. Several people asked for answers about why this was happening. They also wanted information about the role of the refugees. There was a real sadness in the air caused by this situation. The Group seemed very anxious and could not wait for me to start the channel.

As we began, a lady in the front row had tears in her eyes from the moment the Group said "Greetings from Home." She had brought her friend to translate, but shortly after the channel commenced the translator stopped speaking. Afterwards, they both approached me. It turns out that this dear Master Healer did not speak a word of English. With tears in both our eyes she made a gesture of putting her hand over her heart. Her friend, the translator, then told me that she understood every word on a much higher level.

The Group:

Greetings from Home.

We have waited a long time to say that. We have waited a very long time to visit with you. We have waited for you to find each other and re-unite. We thank you for reaching out. We thank you for being who you are. We thank you for stepping into your power and finding the expression of the God within each of you. We are so pleased that you have chosen to walk firmly into plan B and what you are calling Ascension.

From our perspective, we see the turmoil, we see what you consider to be difficult times. We ask you to be very careful where you place your focus, for as you observe difficult times, if you linger too long you create them. There are many events on the Gameboard at this time that do not receive the same attention as the ones that trouble you so. But in order to chase out the dark with the Light one must first shine Light into the darkness. This is now in progress and this is what is being seen in the events unfolding on the Gameboard. Many of you have held the Light in your own various ways. We express a deep gratitude to those who are now playing the role of victim, for they do so in the highest possible manner for the good of all that is. These are contracts they have chosen and agreed to play. These are contracts that were scripted a very long time ago. They are not easy contracts and we honor those fulfilling them.

THE NATURE OF CONTRACTS

We wish to help you understand the events as they are unfolding on the Gameboard. To do this we will first re-mind you about the nature of contracts. Contracts are agreements that you make before you enter the game. You asked this person to come and play the Mother, and this person to come and play Father, and this person to play the part of Uncle Harry. You ask a beloved brother if he will play the part of your business partner. You then ask him if he will love you enough to play the role of the villain. You ask him if he would love you enough to help you learn the lesson and settle the Karma you have chosen for your experience. From a larger perspective, when these contracts are exercised en masse they can easily pull the pendulum of humanity to one side and create possibilities of advancement for the whole. Even so, these contracts are only potential scripts that you have written for yourself. All contracts are contingent upon acceptance. When you come face to face with the contract you are still on the planet of Free Choice and you have choice about whether you wish to carry out that contract. We tell you that there are many on the planet that have chosen to carry out some very important contracts. Our love is so deep for all of you who have chosen this path. The results reach far beyond what you can see in your third-dimensional existence. The new paradigms you are creating through your own evolutionary advances will show in the Universe for a long time to come. The colors that you will wear from being a player on the Gameboard will place you among the highest and you will wear them with pride. You are of this family and we are honored to be here with you.

THE FLOW OF UNIVERSAL ENERGY

We are here with information to help you live and move more comfortably into the higher vibrations of the new planet Earth, which you have begun creating. We re-mind you that it is only information that we offer and we offer it with a deep respect towards your own power. We offer it with the deepest of love. We ask you to take only that which resonates within your own heart and leave the rest, for this is at the base of your power. We also tell you a little bit more about the perspective from afar. Many sub-routines are playing out on the Gameboard. You see the difficulties. You see the dark spots as the Light is shined upon them for perhaps the first time. Your technological advances are bringing communication in ways that you have never had before. With the help of these advances, you are able to shine Light that has never been able to be shined before. We know this is difficult for you, as your perspective is limited to the tragedy and suffering that you see. We thank you for walking forward into these contracts. Please understand that you are not alone in your experiences and contracts. The same connection that we have with you also unites you with all other things. Strengthening those connections and strengthening those energy strands will bring balance and opportunities as you move forward into your own advancement.

We tell you this evening of the reflection as it looks from the Universal energy. For on the planet of Free Choice you not only have choices about which contracts you wish to complete, but you also have choice as to whether you choose to be in the Universal flow of energy or against it. We tell you that from the higher perspective, there is no judgment about your choices. They are simply choices and all choices are honored. There are no good or bad choices. What we tell you is that either you accomplish what you set out to do, or you do not. If you are not accomplishing the creation of your version of Heaven on Earth and experiencing your own passion, then most likely it is simply a basic misdirection of energy. Energy is simply energy, and there is no good or bad energy. It is possible, however, to misdirect energy and to prevent it from completing the natural cycle. This often results in events that you perceive as difficult. We see them simply as misdirected energy that does not emulate the natural flow of Universal Energy.

ERASING THE IMAGINARY LINES

These events happen on an individual basis. They also happen on a community-wide, country-wide and even global basis. You have seen entire countries misdirecting energy. The result is the events you see happening now. We will tell you from our perspective how this process relates to the Universal energy flow. We sincerely hope that this is not too simple for you to grasp. To explain Universal Energy we give you the analogy of the water in your oceans. You may see much of your life on the Gameboard reflected in this water. Observe the movement of the waves as they crash upon the shores and the movement of the tides as they flow in and out. For as the water rises and falls it is balancing the Earth's rotation. View the rising and falling of these tides as simply an act of balance, for the water is balancing the rotation of the Earth. Much the way the waters rise and fall, and the waves come into the shore and retreat, such is the flow of Universal Energy. On the Gameboard of Free Choice you can either place yourself within that flow and ride those waves comfortably, or you can fight that flow and swim against the waves and the tide. Once again, your choices are honored and there is no judgment about these choices. They simply yield different results. Understanding the process can help you make informed choices.

On an individual basis, if energy is directed in accordance with the Universal Energy it will flourish and complete the cycle of creation. If the energy is misdirected, then it will build until it is corrected. If it is not corrected, then it builds and attracts similar energies to form similar misdirections on a community wide basis. If this energy continues to be replicated, it will not be long before it finds expression on a country-wide basis. Similar results will continue occurring and spreading until these misdirections of energy are corrected on an individual basis in significant numbers of people in that country. A full global expression in this manner would result in the destruction of the Earth. This is a conclusion to which you were heading. The misdirections of energy are very basic ones, as energy itself is very basic. The Universal energy is always seeking balance. This is done through blending. Make space for a blending of energies in your co-creations and you will place yourself in the natural flow of Universal Energy. Attempt to segregate or restrict the blending of energies and you will create misdirections of this energy.

(At this point the Group paused in silence. After a few moments they explained why with the following.)

Someone in the room has asked for a healing. We pause in this moment and ask that you also reflect your energies towards this person and allow that to be. (Pause)

And so it is.

The Universal flow is predominantly reflected in many of the things that you call nature. Placing yourself within the Universal flow allows you to achieve your balance very quickly. Yet many resist this balance. Much like the waves we have spoken of, if you were to take a drop of dye and place it in the water at the ocean's edge you might be able to look into the water for a moment and see the color of that dye and enjoy the vibration of that color in its purest sense. And yet, it would not be long before the waves would come crashing in and mix the dye with the rest of the colors in the ocean. From your perspective, the beautiful color is now forever lost to the vastness of the ocean. From a higher perspective, we tell you that the color of the ocean, as viewed from other planets, has a rich vibrant hue. This beauty is only possible because of the many drops that make up the whole. We tell you that this blending is appropriate, for it truly is the Universe seeking balance. Understand that your perspective determines the reality.

BLENDING EMULATES THE UNIVERSAL ENERGY

The blending of energies is the natural order of the Universe. For this reason, you see that people, governments, organizations and businesses that oppose the blending process are destined to experience misdirections of energy. It is these same misdirections of energy that are causing so much turmoil on your Gameboard at this time. There are leaders on the Gameboard now who are choosing to segregate the dye and to keep it within a confined area. To prevent the colored water from blending is against the natural flow of Universal Energy. In truth, all water is energetically connected to itself, just as all of you are connected with each other. Holding these waters apart is only done at great effort and causes great strife on the Gameboard.

A part of the Game now in progress is only possible because of the choices made by the players. Even though the scripts have been written the players have a choice as to what roles they will accept. To accomplish this lesson of humanity on a global scale many have chosen to accept the role of victim. We tell you that the colors of honor they will wear for accepting those roles will remain with them forever. They will always be known as the Precursors of the Light. For they are able to shine Light in a fashion that very few can. And for this we are thankful.

Walking in your power is equivalent to placing yourself within the natural flow and ebb of the Universal current. You are beginning to see glimpses of who you are and how to use this power. You have come here and agreed to play the game behind veils that keep you from knowing your true magnificence and it is difficult to see how powerful you are. We deeply honor you for this. For by going within and clearing the paths to Light within yourselves, you also set the energy on a much larger scale. We ask you to keep in mind that you are making headway. Much progress has been made in the last few months. Although headlines speak of turmoil, please understand that in your field of contrast it is necessary to have dark times to truly see the light. For this is when the greatest Lightwork can be done.

THE CUTTING EDGE OF CHANGE

These opportunities lie at your own doorstep now. You have done well, you have chosen, it is in motion. This family that has reunited this evening has been here many times before. We tell you that in this room many have chosen contracts that have been very difficult. Those at the cutting edge always experience the most friction. And we thank you for taking that role. Your choices in this area have facilitated opportunities to change the paradigm of all that is to come. You see yourselves as citizens of an area, citizens of a community, citizens of a country. We tell you that it will not be long before you view yourselves as citizens of the Grand Planet Earth. Beyond that, there will be a time when you see yourselves as Universal citizens, much the way we see you. Right now, you look to your neighbors and you see the differences between you as cause for war. Yet we tell you that your perspective will change once you see that you are not alone in the Universe. As more is known, as some of your own heritage is starting to be revealed, you will understand more about your own nature. You will truly become citizens of the new planet Earth.

WHY GLOBAL WAR IS IMPOSSIBLE

This blending has begun in many ways already. We wish to point to some of these, for we tell you that a global war is not possible. You are simply too intertwined and, in many ways, you would be bombing yourselves. Part of the way that this is reflected on your Gameboard is through your own economic structures, even where you have chosen to combine your economic structures in the form of what you call a "euro-dollar." And although there is typical resistance to such changes, we tell you that these are the co-creations that are clearly aligned with the Universal Energy. They encourage blending, they encourage moving from a field of polarity, or segregation, into a field of unity. We have spent much time in our writings telling you to please learn how to center your own energy, for it is important to understand that each one of you is the center of your own universe. It is important that you check your own motivations within yourself and that   you use your own discernment as it resonates with your own heart, rather than those around you. These actions, together with learning to speak your truth, are important tools that you will be using more and more. Yet this confuses some of you because you appear to have the notion that this is selfish. Let us illustrate the difference between selfish and self-first. Once you place yourself first in the flow of energy, it is then possible to connect with all other things. Upon connecting to these other things within your field you will achieve balance. You will find that you are all one. Much the way you have defined your territories, your governments, your religions, and your belief systems on the planet, we tell you that these are imaginary lines that do not exist. What really exists is the gradual flow of energy emulating the waves crashing upon the shore. We ask you to open the barriers between these imaginary lines. Allow the energy to flow and seek its own balance.

Strengthening your connection to all that is allows you to be nurtured by the Universal Energy as it passes through you. This can only be accomplished by placing yourself first in line to receive this flow. This takes courage, for you were taught always to put others first. There is an important distinction that we wish to make here. Placing yourself first in the flow of energy and cutting others off is what you would term selfish. Placing yourself first in the flow of energy and then using that energy to feed others is self-first. You have nothing to give from an empty cup. By placing yourself in the Universal Flow you have much more to give other people because you fill your own cup first. As in all actions on the Gameboard, if you can place yourself in the natural flow of Universal Energy your ride through life will be smooth and effortless. Placing yourself in opposition to this energy will attract resistance to your every move.

This applies on more levels than you are aware of. It is in your nature to create sub-routines that materialize as your governments and organizations. These sub-routines are a collection of the overall vibration of the individuals that make up the organization. Seek to direct the purpose of these organizations through your own intent and thereby align that intent with the Universal Flow of Energy.

Take responsibility for creating your own environment, and if the environment is not to your liking have the courage to choose again. We know this is difficult. We know the veils are firmly in place. We know that you cannot see who you really are. We ask you to look into each other's eyes, for this is where you will see us. This is how you will most easily re-member who you are. Have the courage to stand firmly in your truth.

We are with you always. You are never alone. You have far more guidance than you can perceive. The power of the eyes that watch your every move is far beyond your understanding. We know there are times ahead when you will experience difficulties. If you find yourself losing balance, please remember to reach out and offer a hand to the one next to you, for in doing so you will be strengthening your connection to All That Is. Do this often.

Because of the choices you have made your Game has now moved fully into Plan B. We honor you for these choices and we love you beyond your understanding. For eons to come there will be many who visit the planet to see how the Gameboard of Free Choice came to its highest conclusion. We are honored to have you represent us in this fashion. The seeds that you have planted are good seeds. You have done well. You do not always see the fruit of these seeds, for they rarely make the headlines. But we tell you that none of this would have been possible had you not accepted your contracts to play this Game. If you look around you, you will see the evidence. Globally, your crime rates are decreasing. You have becoming a gentler people. You are moving from a motivation of survival to a motivation of unity. As you move toward that Unity your highest purpose will become apparent. Please do not be discouraged as the steps are right in front of you. Know that we love you and are with you always. We respect your choices and we honor your Game.

And now we will take questions.

LIVE QUESTIONS FOR THE GROUP II

At all seminars, after the live channel, the Group loves to answer direct questions. Here are just a few from this seminar:

HOW LONG WILL IT TAKE TO RE-MEMBER?

Question: How long will it take to find out who we are?

Answer: *You are beginning to see glimpses of this now. You will begin to understand more of your true heritage in the near future. It is very difficult for us to give you time lines, for you are constantly changing these. Your own advances create the future as a moving target. We will tell you that, as it now stands, you will become more comfortable within the next six months. You will be personally visited by spirit. This visitation will bring many answers to the questions that you have. You have done much to earn them.*

WHEN ARE OUR CONTRACTS COMPLETE?

Question: How do we know that our contracts are fulfilled?

Answer: *That is a very good question, for you do not always know that your contracts are fulfilled. Contracts are choices. All contracts are simply contingent choices; contingent upon your choosing to accept them. Sometimes contracts are very simple and consist of only a well-placed word. Sometimes they are a simple pat on the back or an encouraging smile. Sometimes they involve a lifetime of support.*

Changes within your DNA have made you extremely sensitive to Universal Energy. Use that sensitivity to discern if your contracts are complete. If there's more to do, be about it. Also, have the courage to align your energy with the Universal flow, for when you are centered in this manner your contracts will continue to add to your life. If what you are doing is constantly pulling you out of balance, chances are you are hanging on after that contract has been completed. *Imagine yourself as a spinning top that is seeking to find its balance. If the weight of your contracts is evenly distributed then your top will spin smoothly and evenly. Too often you refuse to let go and hang on to your contracts long after they have been fulfilled, causing your top to spin out of control. If you check your balance often it will be your best indicator of when to let go.*

WHAT ABOUT KOSOVO?

Question: With regard to the events currently taking place in Kosovo, you say that they are all playing a game. Does this mean we should just stand by and let the game play out?

Answer: *No, it does not mean that. Your courage has surfaced because you have chosen to not allow the pendulum to be pulled any further. To stand by and let it play out would be to not play the game at all and let it unfold by default. Your purposeful step is to create your own reality and all actions should reflect that.*

You are moving very quickly into an environment in which the word "fight" will no longer be used. It will simply no longer be necessary. Yet that is very difficult for you to understand at this point. When you are connected completely to your higher self you will also be connected to each other. When each one of you understands that you are all the same, then it will be easier to act as neighbors and support each other, even allowing for your differences.

When you speak of your contracts, your greatest contract is to center your own energy and learn to create your own reality. In the third-dimensional world there may be times when it is important to speak your truth forcefully. This is part of the pendulum that is in motion. It would not be moving if you allowed it to simply be held to one side. We ask you to create your own reality by choice, rather than by default. Center your energy within yourself, for that is when you make the most effective choices. Fear not of making the wrong choice for that is not possible. Consider all feedback, and keep in mind that if you do not like your reality you may choose again. Did that answer your question?

Yes, thank you for clarifying it for us.

How can I follow my Heart's Desire when I have Bills to Pay?

Question: If my true work is my heart's desire, then how do I do it when I still have to pay my bills?

Answer: *Difficulties arise quite often in the areas that you call work. Our perspective of what your work is, is probably quite a bit different than yours. For the sake of semantics, let us say that what you term is what you do for a living. Quite often, a healer like yourself will take a job or a career to achieve balance and find expression in other areas. This works well for the most part, unless that job consistently drains your energy. On the Gameboard, you have written the rules by which you play the game. In the higher vibrations, we now ask you to find the courage to rewrite the rules. So many people are finding that they are no longer a match for what they do for a living. And it is not the job that has changed, it is them.*

As you advance to higher vibrations your work must also advance. If it is possible to change the work you do to fulfill your contracts then you have scripted your roles well. If you find that you are no longer a match for your work you have options. One option is to find balance by finding groups of like vibration and connecting with spiritual families. By balancing in these areas it helps the higher vibrational person to balance and find expression in their hours away from work. If your work is not a match for you it will only be a short time before you must part ways. In the higher vibrations to which you are ascending, your success will be directly proportional to the amount of passion and joy you experience on a daily basis. *If your job does not provide for that passion and joy, then you would be wise to change to one that does. Your path will never be found in a job that you tolerate just to pay your bills. Find your passion and go after it fearlessly, for that is when you will be able to find your true path.*

For now, we ask you to keep your balance as best you can. If what you do for a living is a constant drain on your energy then you have only two choices. You may change your job or it will change you. Most of the time this will show in your own biology revolting, causing illness to tell you that it is time to remove that energy drain and move on. If, however, you are still somewhat of a match, or you are still fed by some parts of your job, there may still be contracts to work through here, in which case balance can be attained in other areas of your life. There may be important things still to be accomplished. Ask that these be shown to you. Know that you are the most important person in that environment and that you must receive compensation, you must receive an opportunity to express the God within you at all times, in all relationships. If you are unable to do so at work, find other places to express this until you can find other places to work.

BECOMING COMFORTABLE WITH ENERGY SENSITIVITY

Question: I have a question regarding Reiki. I'm finding it sometimes difficult to manage my energy from turning on or off at certain times.

Answer: *The energy movements that you are personally experiencing have to do with your own DNA changing. You are extremely sensitive to energy and are what we call an empathic sensitive. You pick up emotional energy and are sometimes not aware that it is not your own. This condition is commonplace as you move into the vibrational advances, as you move into your next evolutionary step.*

The next thing that we ask you to keep in mind is that so many of you are in the process of re-membering. There are many master healers in this room. A master healer is simply a person who has mastered the art of healing in this or a previous lifetime. The art of healing itself is about creating space for other people to feel comfortable enough to heal themselves. This is the only type of healing that is available Universally. Your expression of it should not be limited to one modality. The cosmic humor here is that you have created modalities of your own that you have not fully re-membered. Allow yourself to stretch. When your teachers say to place your hands in these locations and run the energy between them, and your heart says "no, it's going to work better if I move my other hand over here," allow yourself the freedom to speak your own truth. Allow yourself to find what you know to be true and you will find balance.

As a sensitive, we ask you to not be alarmed at the seemingly uncontrolled energy. There are many vibrational areas that are moving all at once. Your own vibrational advances, the advances of the planet, the advances of your own community, all make it difficult for you to tell what is your problem and what is theirs. The challenge that each one faces sometimes gets blurred by the challenges of the others. Be patient with yourself. Ask for guidance. You know within your heart. Follow it. Your heart knows the way. Did we answer your question?

Yes thank you very much, and thank you for being here with us.

SYNCHRONICITY EXPLAINED

Question: Will you explain synchronicity?

Answer: Synchronicity can be explained on many levels. First, let us say that we often speak of it as a lifestyle to which you are learning to become more accustomed. We have described it as walking down the linear hallway of time. You are moving from a linear hallway to a circular hallway, or a "now' time frame. Part of this is encouraging a synchronistic lifestyle, because to create the type of lifestyle that has worked for you in the past simply no longer works the way it used to. It is simply no longer being supported. It is you that have changed, it is you that has raised your vibration. A synchronistic life style has to do with following your path, pushing gently at the doors and seeing which ones open, and having the courage to walk past the fears and push on them in the first place. Then, once they open, having the courage to walk in. About the synchronistic lifestyles, the best that we can tell you is that it takes practice. Begin at small, comfortable levels. Begin with small easy things instead of trying to adopt all at once. Resist your human urge to change everything in one blink of an eye. Allow us to work with you.

There is one anomaly we wish to speak of, for this will affect many in the room. That is what we have termed as side doors. Quite often, you walk down the hallway and push on the doors and the doors don't open. After you become discouraged at beating on them for a time, you sometimes move on to find other doors. Pushing against another door, it flies open and you look inside and you say, "this is not where I'm going, this is not the door that I have chosen for myself." You think about moving on down the hall and pushing on other doors. We ask you that when the synchronicities line up, take these as a sign from your higher self. When you open a door to a room, and say, "this is not where I'm going," at least have the courage to step into it, for as you take the singular step into the room you will often look to your right or your left and see doors that lead into other rooms that you would not have been able to see had you not had the courage to take that first step through the door.

We often talk about co-incidence and synchronicity as one. When you set about a co-creation, ask spirit to bring you the highest and best, ask spirit to put you in your contract, and then release spirit to work in your life. This is what we call co-creation, for you *are working together with spirit to create your environment and your version of heaven on Earth. Once your co-creation is released, spirit then begins to line things up for you. Spirit moves this person over here, they line up this job, these opportunities. They reach a hand into your three- dimensional world and move things around. This quite often appears to you as coincidence. Please keep in mind that there are no coincidences. Synchronicity is a way for your higher self to speak to you directly in your daily life. Have we answered your question?*

Yes, and thank you so much for validating my thoughts.

SIMPLIFYING SCIENCE

[This next question was asked by a scientist from the Atomic Energy Agency. Most of us in the room did not understand either the question or the answer on a conscious level, but I include it here because the Group feels there are several who will. The illustration referred to can be found in chapter 7 on Co-Creation.]

Question: If you take the illustration of God that you have just shown us, and extrude it three- dimensionally, can you then explain the relation of synchronicity to the three dimensional representation of the illustration?

Answer: *There is no direct answer to your question as it is not as complicated as your question would make it seem. Understand that you are more than can be contained within your bubble of biology. The remainder is what we call your higher self, and synchronicity is simply the means by which this higher self now speaks to you. We can explain it in terms that you will understand, and yet to do so would not be for the highest good. The easiest way to explain this is to reiterate that, by adopting a synchronistic life style and leaving room for spirit to work in your life, you will most easily create your highest good. That is the way it is. You are honored, we* do not mean to demean you in any way, for the scientists are very important. *We will say that there are some very basic misunderstandings at the base of your question. To answer the question, let us simply say that there is order in what you call chaos. Soon, the architects of the energy, whom you have called aliens, will be returning to visit this planet. They will have a very important role to play in redesigning the energy to support the higher vibrations on the Gameboard. You will understand our answer more fully as this unfolds.*

Scientist: Does that mean that the illustration that you have put on the board does not equate to three dimensions?

Answer: *That is correct. The illustration is a simple one, and yet, even this is more complicated than the concept it represents. Understand this, for us to reach into your vibration, we must produce something that is easily understood by your minds. However, in order to accomplish this we must complexify the concept enough for you to understand. It is not possible for us to draw in three dimensions on a pad. Please keep in mind also that there are many more than three of what you call dimensions.*

WHAT ABOUT THE FOURTH DIMENSION?

Question: Can you give some information on the fourth dimension?

Answer: *It is easier for us to explain the fifth dimension to you than it is to explain the fourth. This is because the fourth dimension is not clearly definable in your reality. Let us explain it in this fashion. The third dimension, in which you currently reside, is a resting place wherein you play your Game. The fourth dimension is a place to travel through; It is an interim between the third and the fifth. The sixth and eighth dimensions have similar attributes. These are the dimensional levels that allow you to move from one to another. As* *you travel through these, they imbue you with impressions, or attributes, which you carry forward into the next resting dimension. Beyond the eighth dimension this pattern changes.*

Much the way it would be very difficult to teach a high school class in kindergarten, so too is it very difficult for us to use terms that would be easily understood. Let us explain it in this fashion. Your senses are beginning to attune to other dimensions. Your eyes, ears, nose, mouth, and your sense of touch are the manner in which you interact with the third-dimension. You are starting to assimilate energy in ways that you never have before. This growth is now happening at an exponential rate right at this moment. These changes are causing some stress within your emotional bodies. Part of what is happening is that you are beginning to understand and see beyond the ranges of vibration that you have previously been able to perceive. So, much like a dog would be able to smell things that you would not be able to smell and see things in ranges of vibration that you would not be able to see, your own senses also are beginning to expand.

As you accept this, you will be able very clearly to see the fourth-dimension as a place to simply travel through. The illustration that we have used prior is that of the bridge. The fourth dimension is a bridge that you are building with your own vibrations between the third and the fifth dimensions. The interesting part is that there are some that will be skipping ahead. There are many that are coming to the planet very soon that will be working in multiple dimensions at one time. Although this is possible, there are very few that are doing this at this time. The children of crystal vibration will be bringing the seeds of a new humanity that will make it possible to walk every step. For even now, you often walk in the first dimension, totally unaware that there are tools in that first dimension that you can be using to affect your third. Please be patient with yourselves, and we ask you not to reach too far for the answers, for they are closer than you think. They are simpler than you think. Allow them to find you naturally and experience them, and do not be afraid to reach out and explore those other dimensions. The rules of your game will appear differently in each one, yet you will adapt very quickly. Did that answer your question?

Yes, and I wish to thank you all for this talk, it has answered a lot of questions for me.

SPEEDING UP THE PROCESS OF RE-MEMBERING

Question: What can we do to speed up the process of taking our power and re-membering?

Answer: *The best way to speed up the process of re-membering is to reach out and connect with spiritual family. Connect with others of like vibration. Look through their eyes. Experience who they are, and re-member parts of yourself. In doing so, they will reflect ways that will help you to re-member and accept your own heritage. The ego has been an important part of the biological process. As the Earth cooled you had to take on denser bodies so that you were able to interact with the Mother as she gained density. At that point, it was necessary to incorporate the ego as a survival mechanism. You are now at the point of releasing the need for the ego. And although you are certainly not there yet, you are moving in that direction for the first time. As the ego releases more, you will be seeing more, you will be seeing more of yourself. If we were to show you exactly who you are, and what you had done in previous times to set up these contracts, and how many lives you had moved towards the Light, your ego would not let you accept that information. We are working in an area where the ego is beginning to release. It is difficult for you to accept your own magnificence, yet if you will look through the eyes of your spiritual family you will have no choice, for the truth can easily be found there. And as you look through the* *eyes of your spiritual family you will see your magnificence, not as a singular event within you, but as part of the whole and part of the connection that connects all. So for now, seek them out. Find groups of like vibration and similar beliefs and discuss and interact. Stay with these people and reach out to them. In this fashion you will clear your own emotional restrictions to carry more light through your own biology. This is what we call Lightwork.*

We honor your process on the Gameboard. You have no idea what it looks like from our perspective. If you could only see yourself for a moment as we see you, you would never again doubt yourself. You would move into joy and stay there. You would move fully into your own passion and enjoy every moment of the Game. We tell you that this passion is the same passion that we experience all the time on this side of the veil. When you find your passion and allow it to be a part of daily existence, you are, in fact, creating Home on your side of the veil. When you create Home on your side of the veil, like vibrations attract and the separation between the worlds diminishes.

Interaction is beginning to be possible in more ways than you have ever thought. Stretch your limits. Find your power. Open your own channels. Ask, and it shall be given. Speak your truth. There are times when we do not know how you stand it. We see the dichotomies that you face on a daily basis, the things that you have to view, the atrocities that you have to endure. Yet we tell you that it is only possible to move energy from this position and you are loved beyond your imagination for having the courage to walk within these bubbles of biology. We ask you to reach out and look into the eyes of those around you, to draw your spiritual family together, and re-member who you are. For this is what connects us all. It is our greatest expression to bring information to help you to re-member. It is with the greatest of honor that we ask you to treat each other with respect, nurture one another and play well together.

We are complete... the Group

Chapter 12

The Crystal Grid

Humanity's Connection to the Earth

Live Channel, Elspeet, Holland

Energy Exercise at the Lightworker Spiritual Re-union, Elspeet, Holland

The Crystal Grid

O ur hostess in Holland is Ingrid Kramer. Barbara and I had gotten to know her over the telephone and felt a great family connection, and when we met in person it was even stronger. The event in Holland was a real connection to family and we connected closely with some very special people there. Ingrid ended up accompanying us to all the events we held on our first European tour.

The Lightworker Spiritual Re-Union in Elspeet was a time to re-member. There was joy, tears, laughter and music, and a lot of lives that changed those two days. For Barbara and I, it was like coming home again. This is the live channel that took place at our "Home away from Home."

The Group:

Greetings from Home. You have come a very long way in your lives to meet each other once again in this exact location. Many of you have been here before. It is so joyous for us to see the re-union and to see you look upon each other's faces as you see through each other's eyes. At first, you think you see strangers. We tell you that as you gaze into each other's eyes you see your own reflection very clearly. We thank you for following the gentle nudges that brought you here, for many of you have made movements towards realizing and actualizing your power. We honor you deeply.

Much of the advancement of humankind must be made in human form, for this was the nature of the game. We thank you for doing it, for we know the dichotomy of polarity and the confusion that accompanies it. We know that you get many mixed signals and many confusing illustrations of life. We ask you to simply find your own heart, for as you do, you find us and you return Home. Bringing Home to your side of the veil is the nature of the game, and you have set this into motion. Many of you have gone through very difficult paths to reach this point in your life. Many of you have asked and pleaded to be here in this moment, only to find yourself questioning why you are here. We know the difficulties that you have experienced that have allowed you to be here, and we thank you for following your heart, and we thank you for pushing past your fears and filling that void with the information. We also thank you for carrying the energy of this family so strongly upon the Gameboard. [At this point the tears that had been flowing gently began to overwhelm me].

We take a moment while the Keeper catches his balance. [Pause] Sometimes the energy we have for you overwhelms the Keeper. We wish you to receive the message and feel the feelings directly from Home. In doing so it is easy to hug him a little too tight. [At this point everyone was sniffling].

The tears you experience are what we have termed as life lubricants, for they allow feelings to flow more easily. The restrictions that you have cleared within yourselves have created changes upon the planet that we have not seen in millennia. Many times there have been glimpses of the possibilities to come, only for us to find that humankind was not ready for that to happen. And the Grand Keeper of Time held his finger firmly in the air until the release of the information was appropriate. We tell you that it is you that have made it appropriate. Through clearing yourselves and through clearing and planting the seeds on the planet you have set into motion more than you could possibly know. Your belief system behind the veil would have you think that as you look around the room and see the many things it, this is reality. Many of you are well beyond the understanding of your reality. For you know that it is what you do <u>not</u> see that sometimes is more real than what you do. As your own vibrations begin to change and as the evolution of human biology continues, you may expect very clearly to see things in front of you *that have been there all along. The rational mind was a very important part of your biology and the ego was a very important part of your survival. And yet, you have grown beyond the objective of survival. You have grown and raised your own vibrations beyond survival as the prime motivation. For the motivating fact is now one of carrying light and seeking unity.*

Many of you look at each other and see yourselves in the eyes of those sitting next to you. We know this is difficult for your brains to comprehend, and yet we ask you to simply look and notice the differences within your feelings, for these are more clearly an illustration of your advancement than your thoughts. Matching your thoughts with your feelings will be the very beginning of your mastership. This is what we refer to when we speak of Merlia. For that is a part of you that you have hidden from yourselves for a very long time and we are so grateful to help you re-member. As you begin to reintegrate the previously separated parts of yourselves - the male, the female, the conditional love, and the unconditional love - you begin moving further into unity consciousness. Then you will realize that you were never separate, but have always been one. At that point you will finally comprehend that you are also one with us. The expression of the family of Michael upon the Gameboard is outstanding, for you have done well to bring the love from this family forward. That flow of energy, as it originates, is one of very special balance, one that was needed to find the balance to move from a mode of survival to a mode of carrying light. And you have made it so. We realize that you look at us and you listen to every word that we say with close scrutiny, and still we ask you to understand that it is we that honor you. For yours is the Gameboard of Free Choice. Even though you have choice in all matters, you still chose Light. And we thank you.

We wish to speak of the connections beyond your physical reality, for these occur on many levels. Although we have given information and the Keeper has illustrated the many alternative dimensions and alternative realities that run side by side concurrently, we also wish to tell you that they are not limited to your planet. For it is surely known that life and the expression of **all that is**, is not singularly known on the Planet of Free Choice. There are many Gameboards and many games in progress, yet yours was special. For yours has the greatest possibility of God seeing God. Yours has been that of the mirror, for this was to be the only Planet of Free Choice.

Repaying Universal Karma

At the very beginning of this Game it did not look as if this experiment would continue. At the very beginning, the planet was taken from you, and since the prime directive of Free Choice ruled all, this was allowed as part of the Game. Those who attempted to take the Planet of Free Choice for their own gain thought they would do so for their higher good. We tell you that they ended up playing an important role in a much larger picture. For this was a restriction they had to experience in order to further their own growth. We tell you that many of those have now returned with an honest intent to help, for they are in fact part of your parental race. With their now honest intent to help you move painlessly from one level to the next they work through the repayment of a debt. What you know as Karma is actually prevalent only on the Gameboard of Free Choice, although there is a similar system throughout the Universe. We tell you that there will be very much assistance offered in the very near future from your neighbors. For much is beginning to build on the Gameboard of Free Choice as the energies begin to boil.

The Mother is also Changing

The Earth is fully connected to each of you. Change occurs

within the planet as you make changes within yourselves. As your emotional energies find restrictions and energy rubs that you did not know existed, you begin shining Light in dark places. As you reach to remove these restrictions within yourselves, we tell you that you also release restrictions with the entity you know as Mother Earth. There is a direct energy connection here, and in many ways the Earth energy is only a reflection of the collective human energy on the Gameboard. This is the reason you have averted much of the catastrophic events that you had designed as the end of Plan A.

This energy connection also works in both directions. Be aware that if humanity took a general turn toward darkness, it would not be long before the Earth picked up that vibration and catastrophic Earth changes began again.

Visitors - Your Parental Races

Soon you will be visited once again, and those visitors you will know to be one of your parental races. You will be able to accept them more clearly with the discernment available through knowing your own power. Some of this power is gained through the emotional connection to the Earth. Learn to balance and work with the Gaia energy, and you will be very balanced at a time when that will be of great use to you.

When these visitors begin to make themselves known, we ask you to see them as they are. For in many ways you may feel that they are more advanced than you. In certain aspects they may present themselves to be. We tell you that they are more your brethren than your parents. The interesting part is that they are here for the lessons you will teach them. Hold your power and filter everything through your own discernment. Know well the empowerment factor and hold your own power first. These times could be filled with great joy as they will explain many parts of your history.

THE GRAND LIBRARY

The days will pass after the Earth changes to a full reflection of Home. After the energy returns to a comfortable level and people begin moving even further into their empowerment, the Gameboard Earth will, as it was intended, become the grand library. Beings throughout the Universe, of all types, of all natures, will come to the Planet of Free Choice to see how it was done. For the grand experiment that you have so aptly changed, the grand game that you are at the verge of taking to the next level, will set the paradigms for the rest of the Universe in all that is to come. Never have we had such a grand vision of the collective of God. As you look into each other's eyes, we ask you to please look for the God within. When you find that little spark within others you ignite it within yourselves. Learning to live without judgment is very difficult, for judgment has served you well. We realize that there is much difficulty that you have experienced as you moved forward on your path. We tell you that you are honored for these difficulties, for being vulnerable is at the base of your strength. And we thank you for having the courage to step into your power in this fashion.

ACTIVATION OF THE CRYSTAL GRID

This geographical location [Holland] has much significance, for it is one of the central points of the new magnetic grid. The difficulties experienced on the planet in recent months have had to do not only with your own vibrational advancement, but also that of the Earth. For the Earth is the collective of the whole. And as each individual raises their own vibration it affects the collective, and the collective rises to match the overall vibration. To complicate matters even more, we are at a time when we are re-adjusting the magnetic grids of the planet and something that has been referred to as the Crystalline Grid activation. We have not yet spoken of this. It is time to validate for you that there is a relationship to what you call crystal and what you call Christ. Furthermore, there is a direct grounding of that energy where you now sit. As the magnetic poles shift and the magnetic alignment of the planet changes, the crystalline grid once again becomes active. This is your expression of Home on this side of the veil. Many of you have seen it coming. Many of you have been planting seeds. We tell you that the effects of these seeds reach far beyond what you can see.

If you could see yourself for only a moment as we see you, you would never question yourself. You would never doubt yourselves. You would never ask the question: "Am I worthy?" For you would know that you carry with you the vibration of Michael. And we ask you to carry this and the sword of Michael with you as well, for it will serve you to step into your empowerment. More will be brought forward about the activation of the crystalline grid. We ask you to pay close attention to those words, for there is very deep meaning. You will receive many cosmic winks as a result of listening and learning to pay attention to the synchronicities of life on the Gameboard. As you do you will find the words crystal and crystalline to have new meaning. You have brought yourselves to sit in this room, to connect with one another with your big hugs and gentle nudges. We know that it has been a stretch for many. We know the difficulties that you have experienced to bring yourself to this point, for many of you have had to experience much turmoil in your life. Many of you chose contracts that were difficult to carry. Many of you chose weights and burdens to carry through your life, and we tell you it was only possible because of your willingness to carry these weights. It was only possible because you chose to carry light. You are here for a reason, and it is we who are honored to be in this room with you. And we thank you.

CHOSEN ROLES

Many of you in this room have chosen very special roles of carrying light. *Many of you have chosen to experience even more turmoil. We ask you to remember that you are not alone, that we are here for you always, that we are part of you. Many of you have direct connections to us and we thank you for that expression. We offer you information for the heart. We offer you information to help you find your own connections, and we thank you for asking.*

There are many questions being held at this moment, please feel free to ask at this time.

LIVE QUESTIONS TO THE GROUP III

At all seminars, after the live channel, the Group loves to answer direct questions. Here are just a few from this seminar:

RE-CONNECTING THE ENERGY IN KOSOVO

Question: What can we do about the situation in Kosovo? What can we do to most effectively re-connect the energies there?

Answer: *There is only one solution to disconnected energy, and that is to re-connect it again. There are many who find themselves in stalemate positions, or blocked into a corner from which they feel they cannot remove themselves. Provide the space for them to balance with love and with understanding and the situation will quickly resolve itself. Provide them with a way to win and all will win. We ask you to understand that, as drastic as the situation has been, it has also served a wonderful purpose. These are the last vestiges of resistance to the natural energy within the Universe. For as the Universal Energy moves to seek balance, and as the Universe facilitates the blending of all races, of all energies, that is when you will see the true face of God. You see yourselves as individual sparks of the whole and you think that to hold that individual spark within yourselves is the answer. We tell you it is the blending process that puts you back in touch with your power. Yet there are those who* *find themselves in resistance. There are those who fall into fear. They grasp for the known and familiar. They grasp for the way they have always done things. This has worked very well for them in the lower vibrations of the planet, butt is appropriate that this not be tolerated in the higher vibrations, for each one is connected to the other and it is not appropriate to allow the mistreatment of humanity in any form.*

If you would see your neighbors as the part of God that they are, you would ever treat others poorly again. For as you feed each other and nurture each other, you nurture yourself. Those who have played the role of victim in this script have done a marvelous job. And these are the ones we spoke of when we said; "the meek shall inherit the Earth," for they play roles through their vulnerability and their suffering that will take this light and spread it throughout the Universe. The pendulum has been pulled so far to one side that it will never have to return that far again. We ask you to make space for these people. Send them love. Send them energy. Allow space for the government to save face while making corrections. For to end war with war only propagates more war. Yet it is necessary at the same time to stand firmly in the truth and to stand for what you have termed the rights of humanity. For these are honored. This is seen from this side as God protecting God. Honoring each other as you honor yourselves is to honor the God within you.

There is a further message for you. You are totally unaware of the many lives you have touched. You are totally unaware of the seeds that you personally have planted in this lifetime. We wish it were possible to show you that, much like a spider web, you touch lives that then go out and touch many others. We wish to thank you for. You have done well and we are very honored to be with you.

Understanding the Dangers of the H.A.A.R.P. Project

Question: Can you tell us about the H.A.A.R.P. project?

Answer: *We find great humor that you play with yourselves in this manner. For what you try to do is to take scientific knowledge and twist it and play with it in such a way that would enable you to become more powerful. The dichotomy and the humor of this situation is that in many instances all you do is cripple yourselves. To bounce your own waves off the ionosphere creates ripples that will continue for a very long time. Much of what you have done and much of the process you are experimenting with at this time has to do with using the ionosphere as a parabolic mirror to reflect these waves back into the Earth. There is much truth that has been brought to the planet through what you have termed as scalar waves. These are multiple waves that compound upon each other and have the ability to travel through elements that cannot be penetrated by waves in their normal singular or dual-wave forms. Please be patient for we are planting seeds. There are two people in this room listening to this information, and a third person who has yet to hear it, for whom this will be particularly important.*

Your ionosphere has a natural curve that resembles a parabolic mirror. As these waves reflect back from the ionosphere they focus with intensity on a singular spot deep within the Earth. This changes the Earth's vibration in ways that can be measured. This action is similar to the way in which vibrational healing affects your body. By introducing vibrations of a specific nature into your body you can effect healing by displacing vibrations of ill health. The main difference between the actions of the HAARP project and vibrational healing is that the Earth did not request this healing. Also, the HAARP project is not being conducted with healthy intent. Now, imagine spinning yourself around time and time again and then trying to walk a straight line. In the same way that this would affect your own balance, so too does it affect the balance of Mother Nature and the Gaia, of whom are very much a part of you. Your governments have seen the confusing effect this energy has on the human mind. What they do not take into account is that this also confuses the Mother. It sets things in motion, such as earthquakes and volcanic eruptions. Because of your advancement you have managed to avert many of the harmful effects thus far. Now, to override these energies with your own emotional and spiritual advances and then turn around and aggravate the situation is a clear misdirection of energy.

There is no way that mankind can forget what it has learned, nor do we ask you to. We do not ask you to stop pursuing this technology, rather we ask you to slow down and redirect your focus and intent. For in your quest to build these weapons of war you have actually gained valuable knowledge that could be used for great good. It will help you reveal truths about yourself and your planet. Again, we use the analogy of the HAARP technology and vibrational healing, for it can be yours to use this technology to heal the Earth in the future. Focus this technology for energy transmission instead of tools of war. We ask you to breathe life and breathe love into the direction of these experiments. Does that answer your question?

UTILIZING YOUR POWER TO CHANGE THINGS

Question: So we can change the process by breathing life and love into it?

Answer: First, become aware. First and foremost, bring awareness of a problem to the forefront of your consciousness and of those around you. Speak your truth, write about it, share your concerns with another. And as humanity integrates these new technologies, you will discover there are positive applications. Humanity carries much more of the Light energy than that of war and destruction. The natural process is to find balance. Mankind will actually use this *technology for this purpose. So, first and foremost, ask your leaders; "what is this for?" Bring light and attention into this situation and into these questions. Make these questions public. Present the information to each other and then make the decisions. Much of what has been termed as harmful to the planet has already happened. For these experiments have been going on for a very long time. In some ways, the Earth has even built up a resistance to it. Yet, it has impeded the overall vibrational raising of the Earth. And we simply ask you, not to do that. Point the use of this information and technology in another direction, for it will serve you well.*

THE CRITICAL ROLE OF CHILDREN IN THE NEW ENERGY

Question: What can we do to further our own advancement and the creation of Home on this side of the veil.

Answer: *This is a subject very dear to our hearts. You have personally chosen to be at the cutting edge and experience the difficulties of changing the energy. And simply the fact that you would ask such a question shows your progress. And we thank you for taking your role and you will find more. The doors will begin opening in many areas for you personally and for those of similar vibration.*

The systems that you know as your schools will begin to change. We ask you to be patient, for many of those that have come to make these changes are not fully in place at this time. Many of those on the planet as children have no place to fit into as yet. They feel as though they are visitors here, waiting for the rest of humanity to catch up and it is difficult. Some of these are what you call the Indigo children. They came in as SYSTEM BUSTERS and they came in to rearrange your paradigms for you. It is a difficult process. To make it safe for them, to let them know that it is normal for them to know who they are, is one of the greatest gifts you can give mankind as a whole. Much will be brought forward as a result of these beings on your planet.

Those we have termed Crystal children will move forward to take their place only as a result of the rearranging caused by the Indigos who are now on the planet. Traditional methods of child rearing will not work with the Indigos. They do not respond to such things as guilt. Many times we have talked about the cords that connect you to one another, the tubes that carry unconditional love. Many of you have learned to yank very hard on each others' chords. Yet to yank on the cord of a child of indigo vibration would not have any effect. They move into the different paradigms, not only of your schools but also of your businesses and your governments. As they move into their leadership roles, they help to make it even safer for those to come forward of crystal vibration. For that is the full activation of the Crystal grid of which we spoke. And it is underway. In the meantime, we ask you to examine your own choices, your own belief systems that limit you in the raising of your children, in the teaching of your schools. Examine everything as it relates to this new breed of children, for they are leading the way. They are helping to rearrange your world, as you have asked them to. They are an expression of yourself and they are here with great love. They are very special children.

How can I know I am on the right path?

Question: How can I know I am on the right path?

Answer: There is another expression of this within this room. You have found it in the words in a magazine, for in fact it is a mirror image and a cosmic wink to simply let you know that you are being watched by the God within yourself. [They were speaking about a magazine that often runs the Beacons of Light in Dutch. It is "SpiegelBeeld" magazine, which means "Mirror Image" in English.] It is the grand mirror, and looking in that mirror is the only way you will know if you are on your path. As you begin to become more aware you will see more of these cosmic winks that will tell you that you are on your path. These winks let you know that we see you and we know you and we love you.

The crystals you have placed on your path will also be your guides. These bring the feelings of joy into your life. Have the courage to follow and accept this joy and watch for these winks, they will lead you well. Thank you for asking.

SENSITIVITY

Question: I would like to know why I am so sensitive to sunlight? When there is one cloud in front of the sun, I lose all my physical strength.

Answer: *Photosynthesis within the bubble of biology is the basis of carrying the light. What you have termed as light is different than the light you receive from the sun, although they are interlaced with each other. For the light energy that we see as light is the highest expression of the love energy in the purest form. And yet, from both sides of the veil, this appears as light in the form of illumination. It is no coincidence that to fully illuminate yourself with knowledge will also allow you to carry more light. The play on words is no accident. In the vibrational stages that you are moving through as your biology changes from one density to another there are harmonics that you experience that make you particularly sensitive to different forms of energy. Most upon the planet at this moment are experiencing extreme sensitivity to emotional energy. You, on the other hand, have an extreme sensitivity to light energy. It is a temporary shift, it simply means that you are in a different stage than most of those around you. The biological changes, as we have helped humanity to understand, will advance and alter in the coming months and years of your time in your expression on the*

Gameboard. Some will experience things that are different than the rest. And this is as it should be. Your sensitivity to light is a result of a change at the DNA level. The entire process of photosynthesis within biology will be rising to new levels soon and you are simply ahead of your time. As you move into your light body, you will no longer have an interaction of light within biology for you will be light and will have an interaction of biology within the light. The change from here to there is what you are currently experiencing. Be patient, treat yourself well. Follow your heart. Follow your inner knowingness, for you are well guided. Once you have made it to the other side of your challenges, please share this with the rest. One of the things that you can do is to allow yourself a balancing within sleep. Your personal sleep patterns have been changing and it has been difficult for you to fully find a balance. Allow yourself time, make space, even if it be in the middle of the day, to find balance through sleep. Treat yourself well.

We will take 2 more questions. There are several in the room that have not been asked.

HOW CAN I FIND MY HEALING ABILITIES?

Question: How can I find my Healing abilities?

Answer: Your healership, your expression, is within you and not within any location. You have been feeling the nudge for a very long time and yet your critical mind has made it difficult for you to follow it. We tell you there will be times when it will be absolutely necessary for you to follow those nudges. For if you stay in the vibrations that no longer support you, your biology will react and create problems to indicate that is time to move forward. Feel the resonance within your heart and choose your life accordingly, for it is not you that makes a job, it is you that makes your life. Your purpose on the Gameboard of Free Choice has always been to reflect *the God within you. Find the place most comfortable to do this and your passion will find an expression at that place, abundance will find you and all the necessary resources will locate you. Have we answered your question? Is there one more?*

Easing the Path for the New Children

Question: How can I help to make life an easier experience for my children than it was for me?

Answer: *There is such a deep connection with you. We listen to your question and what we hear is your honest intent to make life comfortable for those around you. We love you for that. And yet, we tell you that there is only one way that you can make life comfortable for your children. That is for you to move into your passion. For as you do, you give them permission to move into theirs. You are the role model, you are the person they watch. Allow them to stand unhampered, yet assisted when necessary. Let them feel they have support. Know that they stand on their own, but feel the support when necessary. And as they become stronger in standing on their own, reduce the support and move into your own focus. As you center your own energy within you, as you become more, as you accept the energy first in line, we tell you this is when you give the greatest gift to your children. This is also when you give the greatest gift to those around you, for it is not possible to give from an empty cup. Fill your cup first. Create the life that you desire. That will give EVERYTHING to your children. Have we answered your question?*

Yes, thank you very much.

There are two people in this room this weekend that have asked for and have received healing. There are many, many of you here in this room that have opened the door and begun the process of your own healing and have taken matters into your own hands. You have deliberately altered the OUTCOME of your own existence. And in that manner, you have reflected the God within you. We understand that there are times when you feel lost. We understand that there are times when you feel separate from one another and from your connections. We tell you that to do this is for the greatest good of the universe, for this is the expression that you have asked for. This is the manner in which you can change your world one heart at a time beginning with your own. The honor we have to be a part of your process is not describable in words that exist. So instead, we ask you to feel this as we now hug you and bring you re-membrances of Home. As we touch your hearts, you touch ours, and we re-member each other. We leave you with this feeling. It is yours to walk with always. For you are as much a part of us as we are a part of you. The memories of Home are our grandest expression. We will help you to build Home right where you stand. It is with the greatest of honor that we ask you to treat each other with respect, to nurture one another and to play the game well together.

We are complete... the Group

Chapter 13

Home

Re-membering the other side of the veil

Closing ceremonies; the "Angel Walk" at the Scepter of Self Love,
Sudbury, Ontario, Canada

Home

*A*t a San Francisco seminar we met two people who were the embodiment of the work as described by the Group. One was a Master Healer who had realized her calling as what the Group had termed a "Transition Team member." The other was a client and friend who had just begun her own graduation process. She was diagnosed with terminal cancer and was told it would take her life within 6 months. When she heard this news she called me for a session to get clear about her options. With gentle guidance from the Group, we helped her accept a decision that she had already made on a higher level, to leave and return Home. At first she was scared and angry because she had worked hard to prepare for the ascension, only to leave before it was to happen. The Group then delivered a message for her that she was, in fact, to be a big part of the shift that we call the ascension. They said that she was being called Home to take her place and assume her real contract. This was a contract that was not possible for her to complete on this side of the veil. This trip allowed us to put these two people in touch with each other as they stepped into their next contracts together.

Later that same year, we returned to San Francisco for a two-day seminar. During this trip I had an opportunity to re-connect with both of these special people. The transition team member, Treesha, had taken on new life energy. She was literally glowing throughout the seminar. It was obvious that her re-membered passion of working with those in transition was a direct vibrational match for the new person she had become. She was being fed by realizing and moving into her contract, as she had found some very important crystals on her path. Although on this trip Heather was not well enough to attend the gathering, I did get to talk to her. The Group had some very special things to say to her. As I spoke, I was hearing some of these things for the very first time. They said that humanity has already moved to a higher vibration, and because of this movement it was now possible to interact from the other side of the veil more than ever before. They told her that she was to play an active role in what we are calling the ascension. They told her that she had worked very hard in this lifetime to get to the point where she could take her place. Because of her work she now could fulfill her contract from the other side. This has not been an option until recently. Because of the collective vibration of the planet we now have new opportunities.

Two weeks later, we were packing to leave for another seminar when Barbara suddenly stopped what she was doing and said that I should call Heather. I was unable to connect and left on the trip the next morning. Upon returning from New Hampshire we found a message that Heather had graduated. It is my honor to dedicate this message from the Group to Heather. Her work will continue.

The stately elderly gentleman with the white beard has just lowered his hand, motioning that it is now time to release this information.

The Group:

We are in great joy to be here with information at this time in your evolution. It brings even greater joy to see the many ways you are applying this information in your daily lives. We are honored to offer this for your individual discernment and your empowerment. Our job is to help you to fully take your power and re-member who you are. We deeply honor your willingness to allow this information to find expression through your lives. Through your willingness to incorporate this information you have placed us on our path of highest good, and for that we are grateful.

RE-MEMBERING HOME

In this session we wish to speak of something that all of you have experienced many times. On the Gameboard of Free Choice, you express your spirit within bubbles of biology. These bubbles are finite and as such, allow you to take the position of being the finite expression of the infinite Creator. We will take this opportunity to present information that will help you re-member the other side of the veil from which you came. It is time to speak of Home.

There are misconceptions that we wish to help clear up. We have told you that in the near future many will be leaving the planet. This will be for several reasons. Most will choose to leave in order to make a quick return in new biology, carrying the seeds of the new crystal vibration. Technology is now emerging that will allow you to rejuvenate your biology, yet it will not be in place for many years. Even as you begin to incorporate its use, it will still be easier to retard the progression of the aging process than it will be to reverse its effects. For these reasons many will choose to leave and re-enter.

CARRYING ETHEREAL FORMS INTO THE EARTH

At this time, the overall vibrations of the planet continue to rise at a quickening pace. This has an effect to which we have only alluded thus far. You are beings of great capabilities and power. It is your higher purpose to take ethereal forms and ground them into your three- dimensional world. If you observe, you will see that everything you term as man-made is actually an ethereal thought form that you have run through your bubbles of biology to make three-dimensional. First comes the idea, which you attract from the Universal mind shared by all. From there, you take that ethereal thought form and create three-dimensional reality from it. This is the basis of the co-creation process. On an even higher level you may see that your purpose on the planet is to carry the ethereal form of

Light into the Earth. Transmuting Light into the three-dimensional world is most easily achievable through human biology. Because all things on the Earth are connected, the more Light you transmute into the Earth, the higher the collective vibration rises. Raising the vibrations in this manner makes it possible to merge both sides of the veil, bringing Heaven to Earth. This is the highest outcome of the Game.

Because of the changes you have willingly made within yourselves, you have greatly increased the amount of Light you are now carrying into the Earth. This has enabled the activation of what we have called "Plan B". It is this simple action of clearing yourself that enables you to intentionally tune your world to a higher vibration. For this you earn the highest colors and honor of the universe, for it affects more than you can imagine. As a direct result of the vibrational increase, the veil that separates our worlds is much thinner. Also, as a result of the veil thinning, it will be possible in the very near future for interaction between the dimensions known as Heaven and Earth. There is now a permanent corridor of Light manifesting to facilitate this interaction. In the past, all forward vibrational movement could only be accomplished on the Gameboard within a shell of biology. Yet now, as Heaven and Earth come closer together in vibration, it is possible to interact from both sides of the veil. This means that in the next step of your evolution many important roles will be played out from our side of the veil. Many of the grand masters and those holding the energy of this movement on the Gameboard may feel the call to complete their contracts from this side of the veil. This is as it should be.

Glimpses of Home

As the veil thins, many will begin to see and re-member Home. This is already happening. The joy of Home will call to many, and some may decide to leave based on that call alone. The vibrations of Home are powerful and, when felt, they have a pull that is difficult to resist. For validation of this we tell you to speak with those who have experienced the other side through what you call "near death experiences". This experience differs slightly because the cord was never severed, yet these people have peeked through the veil and carry much information about Home. Although they may not fully understand what they witnessed, they have felt the vibrations of home and can tell you how strongly it pulls. In listening to these stories, we ask that you learn to listen to the feelings being expressed, for they contain much more information than do the events themselves. We ask you to re-member your higher purpose, for there was a reason you agreed to play this role on the Gameboard. The attraction of Home may be more evident because of the thinning veil, yet we ask you to see that you have spent much energy placing yourself in the right spot to play the part before you. Your vision is tainted with the effects of polarity and your perceptions may tell you that you are not in a position to help in the grand plan, but we tell you that you are in exactly the right placement for the path you have chosen. To make a conscious choice to return home early would be to leave just before the miracle happens. Do not let your life circumstances discourage you. Instead, use these circumstances to encourage you to claim your power and change your world.

Progression into Plan B will bring great joy and passion, but one must travel the road of change to set this into motion. Moving to higher vibrations within your own being adds to the collective and the planetary vibrations. Therefore, it is a reality when we tell you that you can change your world, one heart at a time. This is at the basis of the co-creation process. As we have described the Grand Game, it is this power of co-creating together with spirit that will facilitate the creation of Home on your side of the veil. This is the way in which, together, we will create Heaven on Earth.

We will speak more of the role that spirit plays so that you may *begin to understand the process more fully. Understanding your role in biology will give you a clearer perspective. You are honored and dearly loved, for through your willingness to play the Game in the lower vibrations you are able to ground these ethereal energies in ways that we cannot. We watch as you experience the pain of being human. We see the deep loneliness you experience, being far from your true Home. We see the confusion at not being able to easily re-member your direction or your power. We see the many mirrors you place in your path to help you re-member, only to view them as circumstances dealing you another blow. It is during these times most of all that we are here for you. Our love for you is deeper than words can express. Often, we wrap you in our loving energy and hug you. Do not ever think that you are alone on your journey. You are the family of Michael.   This is a family of great pride and strength.   It is yours to own your heritage and walk tall in the full knowledge of that power. We tell you once again that it is you that are honored, for through your willingness to play the role on the Gameboard you are setting a space for the healing of the Universe. You truly are the Master Healers of the Universe and soon you will see your own reflection of grandeur.   You are finite parts of the infinite Creator, and as such you have access to all the powers of the Creator.   We are here to re-mind you of your magnificence and your heritage.   Until such time that you fully re-member, if you lose your way and cannot find your path, if you are confused or find yourself in despair, please look at your reflection through our eyes and re-member who you truly are.*

INTERMEDIATE HELPERS CALLED GUIDES

Your role on the Gameboard of grounding ethereal forms into three dimensions allows for our work, for it is we who position those ethereal forms for you to ground.   We will now explain our role further.   In the lower vibrations of Earth it has been difficult to interact with the human form. Previously, for us to interact with your three-dimensional world, it was necessary for us to lower our vibrations. In times past this was accomplished for short periods, but only at important junctures in your development, because of the great expenditure of energy required. It is a Universal constant that like vibrations attract one another. Therefore, it should not surprise you to find that contact with higher vibrational beings was labored and required much effort, for we had to emulate your lower vibrations. It was this space between the worlds that created the need for inter-mediaries to help with your experience on the Gameboard of Free Choice. This has been the purpose of the beings you call guides. These roles are taken and agreed upon ahead of time before incar-nation. Mostly, these are beings that have had experience in the areas in which you wish to work during this phase of the Game. They carry great love for you and often prop you up when you feel lost and cannot re-member your true power. These beings have agreed as part of their forward movement to assume an intermediate vibration that enables them to walk with one foot in both worlds simultaneously. Many of you now on the Gameboard have spent time in the role of service to others. This is a most honored role, for it allows you to give constant love and support to the recipient.

The guides that ride on your shoulder during the Game, are also on their own path of advancement. These wonderful beings are available to you at any moment. Many of you have tried so hard to communicate that you cannot hear the voice that has been in your ear since birth. Once again, we ask you to look within instead of without. Instead of listening for them to talk to you in voices differ-ent than your own, we ask you to pay attention to the gentle nudges that have been with you always. These are presented to you at all junctures of the Game. They originate from the guides on your shoulder and are offered in the deepest of love. They are present because you have asked for them before entering, yet as you play on the Gameboard of Free Choice, it is yours to listen or not. Please understand that there is never any judgment about this choice. This is a source of great guidance, even though it may seem to speak to you in your own voice. Practice the art of listening, for it leads the way to your path of least resistance. Because of the intermediate vibration they have assumed for this journey, they can interact with you on a more physiological level than can those of higher vibration. These are the hugs that you receive when you most need them. If you ask, they will hug you in a way that only they can. From your perspective, it will be felt in both the emotional and physical bodies concurrently. The unconditional love they carry for you is a most wondrous and revitalizing experience. For those of you who have not been able to listen to your guides, we ask you to become more aware of how they touch you emotionally and physically. Allow for them to touch you, and it will be so. They are offering this touch in this moment if you will but only accept it.

THE RETURN OF ANGELS

There is a vibrational hierarchy that we will discuss only slightly, for it is not yet time to release all of this information. Be patient, for it will not be long before this is revealed. Understand simply that there is a hierarchy of vibration, and this enables all things to unfold naturally. The reason we are discussing this hierarchy of vibration is because you have often seen its harbingers. These are the beings you know as Angels. We tell you now that these beings are very real. Many may consider Angels as stories of old, yet we tell you that these are beings of a high order. Although they can only be in one vibration at a time, they are able to move easily between vibratory levels. Those that attain this status carry only the pure light of Home. This is the reason you often see them as white or clear. These beings are often seen with wings that have been described as large bird wings. What you see is presented in a form that you easily understand. The wings you see are representations of parts of their bodies that enable them to travel between vibratory levels and dimensions. These wings are also used to reflect and direct vibration. It is the way in which they affect us from other dimensions. Because of the vibratory level of these messengers, the divine light always surrounds them. There will be much more to come on Angels, as **we** play a very important role in the human development.

It is not by accident that Angels are now beginning to make themselves known to the masses. Much like the seeds that we spoke of in reference to "Light", the seeds of angelic presence are now being planted among humans. The time is drawing near when you will become consciously aware of the many ways you have experienced Angels. Let us simply leave you with the information that there are many types and levels within the angelic vibration. View Angels as a race of the highest order. This is the highest step of evolution prior to re-entering the God-head. Angels are here to show you the highest truth about yourself and to guide you to re-member your power and true nature. This group of nine is a family vibration, each one having their individual field of expertise. So it is with the rest of the angelic realm.

We have spoken before that we are of the family that you would call Michael. Much like a harmonic, this family vibration resonates deep within your own heart even as the word enters your ears. Feel the gentle love, run it through your own discernment and follow the direction of your heart, for it will guide you Home. The love you feel from us and the rest of the angels tells you that we are more than family, we are **your** family.

RELEASING JUDGMENT TO RAISE VIBRATION

As we continue, we ask you to release judgment, for it is important that you understand that judgment does not apply to the level of vibration one carries. The higher vibrations in which we reside is *not any more important than the lower vibratory levels that exist. They are all expressions of God and, as such, are simply levels at which the greater Game is being played out. Your assessment of the higher levels being of greater value comes from your desire to move forward and advance. This is as it should be, yet it leaves you with an understanding that the higher levels are the only ones that have value. Have you ever stopped to consider that there might even be beings who purposely attempted to lower their vibrations? We have just described ourselves in that circumstance, as this was necessary for us to communicate with you in times past. When you look at yourselves through our eyes, do you not see and feel the tremendous love that we have for you? In reality, this is higher vibrational beings looking at lower vibrational beings and seeing the wondrous things that only they can accomplish through their willingness to play the Game in the lower vibration. We tell you that even though our vibration is much higher than yours we are both equal expressions of the Creator. In fact, the circumstances that place you in the lower vibrations enable us to benefit in ways you are not yet aware of. This is the true nature of the Game. More will be seen as you continue your advancement. It is yours to offer vibrational advancement to those choosing it, yet we ask you to judge not your vibration, nor the vibrational level of others. One's vibration only reflects one's position at this moment.*

We have given you an earlier illustration of the ladder of vibrational advancement. It has been your belief that only those of the higher vibratory levels will be a part of the ascension. This is like saying that one can only evolve by way of the top step of the ladder. The ladder of advancement is not fashioned in this manner. In fact, it is designed in such a way as to incorporate the collective vibratory level, and as it reaches critical mass the entire ladder moves to the next level. In your schools on the Gameboard, the third grade is no better than the fourth, they are only stages to move through. Therefore, we ask you to view vibratory levels as something to move through on your path of enlightenment.

JOURNEY TO THE OTHER SIDE

With a clearer understanding of vibration, we now move to the subject of the other side of the veil. We will take you on a journey to give you an idea how this is perceived upon returning Home. Please understand that what we will now share with you is an example of the progression of events that occur when returning Home.

Transition brings about a return of your full power. This power is very subtle and the new inductee is often not aware that they are creating their reality. We tell you now that Home has potentials beyond your experience. You create your individual experiences by calling in a reality that you can understand. Your expectations of the other side determine your first experiences as you arrive Home. This is the reason that many describe this experience differently, and we tell you that all lead to the truth. No matter the experience, it is always woven with the thread of pure love, for in the form of spirit you walk in full union with your higher self. The energy of Love is the predecessor of all other forms of energy. In this place called Home you once again re-member how to live in the most pure forms of Love that we call Light.

THE GREETERS

The first perception is of seeing the "Greeters" that await you to help in the process of transition. They are often seen as beings of Light, or simply felt, but they are always present. Most of the time these are beings you have known that have gone on before you. The Greeters offer their hands to show you that they are there and ready to help. Depending on your beliefs, it is also common to be greeted by angels. At this point, without the proper attitude, it is possible to go into fear and fight the graduation process. If you resist the process, most often you grab hold of something out of fear and anchor your-self to an Earthbound object. *This is the basis of what you have labeled ghosts, or Earthbound entities. Since this process takes place at a time when you are in transition from the Gameboard, the rules of both worlds apply. In all situations Free Choice is yours. With guidance and education this diversion of energy will happen less often. This education is the role of those we have spoken of as Transition Team members. They are now beginning to heed their call and take their positions on the Gameboard. We ask you to support them and make space for them to take their place.*

RITES OF PASSAGE

Envision the process now before you as walking through a tunnel from one vibrational state to the next. There are Greeters hold-ing out their hands in encouragement, yet you are the only one that can walk through. This tunnel is actually the internal experience of the soul leaving the body through the pineal gland. The fear of not knowing what lies ahead can tempt you to turn around or stop walk-ing. For this reason the Greeters often make themselves known to you in your last days of life on the Gameboard. They appear to you while you are still in biology and let you know that they will be wait-ing. If you have been unconscious prior to your transition you will usually regain consciousness momentarily. This is to disengage the connection they have made and allow you to transition. The pas-sage through this tunnel must be of your own volition. No one can walk for you or walk beside you on this journey.

FIRST LIGHT

At some point in this tunnel you will notice the Light. This Light is like none known on Earth, as it is the vibration of Home. It is there to guide you through the tunnel. This Light is all that you will see in the tunnel. *Some are accustomed to seeing Light with their minds instead of their hearts and get confused. If you feel the Light and take it in, you will be easily guided through the tunnel. The hands of your Greeters wait to pull you through the last few steps. The re-union at the end of the tunnel is grand indeed.*

At this moment you find yourself alone being greeted by the Light. The Light is the same pure energy that you have been chan-neling into the Earth, as we discussed earlier. This Light illuminates all the levels of Home. It is the thread of truth that weaves through all things in Heaven and Earth. The Light is truth, and the pure form of the energy we call Love. If you are accustomed to the Light, you will look directly into it and embrace it. If you are unfamiliar with the Light it will be difficult for you to incorporate, and you will resist and look away. Most often, it is the judgments you carry about your-self that make it difficult to look into this Light. If you are able to embrace the Light, let us encourage you to do so for it carries the truth that will allow you to release all judgments. The purity of this Light will instantly re-integrate you to the vibrations of Home by sim-ply allowing it to shine upon you. The degree to which you accept the Light determines the level at which you enter the next dimen-sion. It is important to understand that there is no judgment about the level at which you enter. It is simply a way to direct you to the most efficient starting place. It is this process, which was shown to you a very long time ago, that humans have twisted into the story of Heaven and Hell. Such is not the case, and we tell you that there is no judgment on this side of the veil, other than what you bring with you. Thus the idea of Hell is entirely of your own design. Practicing the release of judgment now, while still in biology, will help you incorporate the vibrations of Home into your daily lives.

REST IN BETWEEN INCARNATIONS

Depending upon the energy conditions of the soul as you return Home, it may be time to rest and heal. If the spirit was very weak and labored greatly during the transition process it will need a period of rest and healing. This process is an important part of returning Home. If a soul reincarnates before receiving ample rest between phases, they will carry with them a tired feeling through most of their experience on the Gameboard. For this reason, it is important to not overwhelm souls upon first arriving. There are also special areas set aside for the purpose of rest. All are encouraged to make full use of these for they help you to re-member the vibrations of Home.

YOUR LIFE REVIEW

You are soon seeing your own life passing in front of you. If energy levels of the soul permit, many will experience this life review at the time of their transition. This is a very personal event and is not what most might expect. Besides being a time to review one's own life experiences, this process serves to indelibly etch these experiences into the Akashic records. Another reason for this assessment is to provide you with an opportunity to decide what you wish to include in the next incarnation. This is usually the time that you are rejoined with your personal guides. The re-union with your guides is a joyous time. Your guides have been at your side every moment of your journey. Now is a joyous time to look at the replays. It is not a time of judgment, for the vibrations of home do not support such actions. Envision yourself viewing all the replays of your life in three-dimensional reality, with your closest friends at your side. This is a time when your guides can tell you of their feelings and perspective at specific events in your life. Often, they wish to reach through the veil and touch your heart to let you know that you are loved, but this is not always possible on the Gameboard. Now is the time when they can share this with you.

There are moments in this review when events in your life are seen with clarity for the first time. Situations that have always stifled you are now made clear. You can hear applause at each important occasion. These important events are not the ones you were expecting. These are the times in your life when you took your power and purposely created your own reality. These were the times when you walked past the illusion of fear to find your true strength. These were times when you were good to yourself, for this was illustrating that you were honoring the part of God that is within. These were times when you went inside and cleared the restrictions you had accumulated; the times when you not only gave love, but also allowed yourself to feel the love given to you. Valuing yourself is one of the most honored acts on the Gameboard. You think to yourself "If only I could have known how this all works, I would have done it much better." The wonderful laughter is heard all around you, and you find yourself joining in as memories return of the last time you said those exact same words. At this moment it is hard to imagine not re-membering who you are. It's actually a great deal of fun watching the replays of a master hiding behind a veil of forgetfulness. At the close of this review you have access to the karmic point system that helps you leave this review with a clear sign as to which direction you will choose next time out. Contrary to what is believed by many on the Gameboard, it is a joyous experience.

MEETING YOUR TUTOR

As the life review closes, you are met by your "Tutor". Tutors are different than the Greeters that met you upon first arriving. A Tutor is one whom you generally have known in biology who has made the transition before you. Tutors are there to help you acclimate to your new surroundings. Quite often, your Tutor is not the one you may have expected. This is for several reasons: The first being that this is done by contract, and at the time the contract is made the full course of a life is rarely known. Often, this is one who has a clear contract to leave before you. These people are often encountered on the Gameboard as those to whom you have unexplained, yet very deep connections. Often this is a cosmic wink. They may be a relative, or a close friend on the Gameboard. No matter how they entered your field, they always had the strong undeniable pull of spiritual family.

Often, this concept of contract is difficult to understand. This is because you attempt to use your human standards. The conception of time being circular is sometimes the most difficult for you to accept. Our dimension of time is much different than your own, and this usually requires adjustment on your part when arriving Home. It is common for those transitioning to experience confusion at this point. To clear this confusion, a time of rest and healing is usually in order.

THE TRAJECTORY OF LEAVING

We will speak now of the importance of alignment upon leaving, for it can easily shorten the time required for this rest and healing. The angle of trajectory with which one leaves the Game, will determine the trajectory of their re-entry. This is the mechanics of what is called Karma. It is also a direct result of the polarity on the Gameboard that produces this effect. Those we call the Transition Team members are honored beings who have dedicated themselves to this purpose. In many ways they are connecting Heaven to Earth in everything they do. Honor these people and make space for them to come forward. They will play a large part in the evolution of humanity.

ATTENDING YOUR OWN 'FUNERAL'

Since arriving Home, your thoughts have been with those you left behind on the Gameboard. Now is a time when your Tutor asks if you wish to attend the ceremony in your honor being held back on the Gameboard. If you choose, you may attend the honoring of your graduation at your memorial. Your guides are at your side for this event, as they were very much a part of your life in every way. They also know and love these same people that were part of this phase of your Game. It is as if they are once again at your side, speaking in your ear and re-minding you of the hearts you have touched. You view these people and often see their pain as they feel the loss. They must now begin the process of redefining their three dimensional lives without you. If it is their choice, they may feel your presence and your embrace. They are most often looking for the undeniable experience of a physical touch. Yet, even if they were to receive it, their ego would intervene and rationalize it into some other form. They feel emptiness at your passing, as their focus tells them that something has been taken from them. What you see is the gift that you have left behind. Each one is now looking at their own existence from a clearer perspective because of this gift. You think to yourself how this event has brought people closer to them-selves, and closer to each other. It is time to leave now, and you look at those you are leaving behind. You feel their pain, and wish it were possible to tell them how wonderful Home really is. You know in your heart that you will see them again soon.

Your Tutor speaks in your ear that now it is also possible to interact from this side of the veil. You can see that it is time for their healing, and this can best be done in your absence. As much as you would love to help them, your Tutor is re-minding you that now is not the time. You will be able to re-connect with them in very short order. Allow them time to find the part of you that is always within them. Then it will be appropriate for you to interact with them again if they choose.

RECEIVING YOUR COLORS

Leaving the celebration on the Gameboard, you are brought to the great assembly in the Hall of Colors. In this hall, all vibration is expressed through color. The colors earned are placed upon you here. The numbers that fill this grand hall are far greater than you imagined. The gathering here is a special occasion. This is to bestow the colors upon one from the Gameboard of Free Choice. This gathering is for you. Now the speaker begins and announces your graduation. Applause fills the great hall as you are presented with colors that you will forever carry. These vibrations, in the form of colors, tell the story of your time on the Gameboard and all your deeds to usher in the new paradigms. Your agreement to separate yourself from your higher aspects for the good of all is one of the most highly honored acts. The colors you are given here will mark you forever as one who felt the pain and helplessness of biology to help ground the energy of Home. Many who see these badges will honor you for being a player on the Gameboard of Free Choice. Your colors also speak that you were on the Gameboard at the time of the great shift. Many times you will be asked to tell your stories of the Game. The feelings of gratitude and honor fill you in a way that makes you glow. Combined with your new added colors, this glow is a wondrous sight indeed! There is no need to concern your-self about ego here, for the natural balance says that as you see your true power you add to the whole. The time has now come for you to leave, and you turn to find your Tutor waiting to take you back home. After this wonderful event you say a temporary farewell to your guides for they are now released with their added vibrations to go on to their next assignment or incarnation.

RE-UNION WITH OLD FRIENDS

Your Tutor was arranged by contract and will remain with you for as long as you choose, answering questions and helping you to understand and re-member the way things work here. You are told that your memories of Home will return, as will all the memories of past incarnations, yet they must be introduced gradually so as to not overwhelm you. During the time to come you often meet many others of similar vibration. Some of them you may even know as your own spiritual family, and some you may have encountered in this last incarnation. Some may have played the part of your Mother or Father, or perhaps a husband or wife. Often, as you are with these people, their features seem to change, evolving from one to another as you speak to them. They are displaying the many times you have known them on the Gameboard. This also happens in your experience while still on the Gameboard. You may find a long lost friend, only to fall into fear as they seem to change into a dreaded enemy you have encountered. The laughter begins, as you understand that you asked this person to play that part because they loved you enough to do it for you. These are joyous times as your memory begins to return. The more you re-member, the more the laughter returns. What a grand Game this was indeed.

STRETCHING OUT IN HEAVEN

Becoming accustomed to moving around in the higher vibrations, you begin to stretch out and use more of your powers. Here in this place there is no time lag and all is possible with thought alone. As you begin to understand the rules of how it works, you may begin to see that these are the same rules we have been giving you for your experience on Earth as the art of Co-Creation. As below, so above. It is not long before you discover the heightened senses that you have in this place. The vibrant colors are like nothing on Earth, as they seem to have an effect on you as they are viewed. The music is of a vibration that seems to gently flow through your essence, leaving a soft, gentle imprint behind as it does. It is as if the music is alive and always talking to you in this pleasant *manner. You are soon told that the wonderful sounds and sights you experience in this place are called in by your own vibration. They are custom made for you and are actually a reflection of your own vibration.*

Even though you do not have a physically dense body you still have the experiences known as your senses. These are simply familiar ways of accepting vibration into your being. These senses are also greatly enhanced, carrying much information. Much like your experience on the Gameboard, each one of these vibrations slightly alter your own vibration. There are many more senses in this place that you have no reference for and, as such, you must discover new ways to experience them. There is one of particular fascination to you. It is the sense of absorption. This is a way of absorbing energy and vibration. An expression of energy gently passes over you offering its flavor of vibration. If you like this flavor, it then incorporates into your being, adding its energy to your own. You become aware that this is also a sense that you had while in biology, yet there was no word to accurately describe it.

Such a wonderful place! Every experience feeds you in some way. You soon see yourself as the clay of the Universe, lending yourself to a shape of your own choosing through that which you call into your experience. You are, in fact, defining yourself in each moment by what you are choosing to experience.

Now, confident of where you are and how this place works, you begin to explore more of the possibilities. Think it, and you are there. Direct your thoughts to an experience and you find yourself in the middle of it. There is so much available here, the most difficult part is in what to choose next. Try your hand at visiting places you always wanted to visit, and you are there. As you open your eyes on each new scene the beauty is almost overwhelming. You cannot help but wonder if you will ever get used to it. You note that the scenery around you is very much like Earth. This is because these are your most recent memories and the way you are expecting it to be. You view the beauty and see it for the truth it contains. Searching for a comparison, you re-member Earth. You ask yourself, "Could it have been that this beauty was always on Earth and I did not see it? No matter, if I return I will know what to look for and I will re-member to look." Your Tutor beams a smile as your contracts come to a close.

CIRCULAR TIME

Here, time takes on a new meaning. Time is circular in this place, and there are new rules that govern the way this works. As you begin to adjust to these new rules you are able to interact with other dimensions of time. The past, present and future are all one big circle, and all are in the now. This is one of the most difficult areas of adjustment, for there was no reference in your most recent experience. A balance is maintained only if you do not lose track of your own energy. Connecting with your energy is the only way to find your place in the shifting dimensions of time. Re-membering the Gameboard, you can see that this truth is also beginning to be revealed. As above, so below.

Things are beginning to return to you naturally as the adjustment settles in, and this feeling of serenity becomes a way of life. You have knowledge of this place that you cannot explain. It is like a memory returning slowly. Without having been re-minded, you are aware that there are many levels to this place we call Home. They are segregated naturally by vibration. The vibratory levels of Home are made up of a dominant vibration, and all the resonant frequencies that lend harmony to this base vibration. Much like a harmonic chord, these vibratory levels find their own boundaries and group together naturally through like vibration attracting like vibration. Together, these make up the levels of Home. When one changes their vibration, they move on to the next level to complete *their work. It is very similar to a graduation and there is much excitement for the one moving to the next level. Movement can be accomplished on this side or on the Gameboard. There are opportunities for great shifting on the Gameboard, and often this can mean jumping several levels at once. There is no judgment about one level being better than another. They all have a purpose and, together, they make up the whole.*

New Levels of Communication

Up until this time, moving to the next level was most easily accomplished while in biology. With the advances and vibrational rise of the Gameboard itself, it was more possible to advance through the levels on both sides. Now, interaction with the Gameboard is possible from this side of the veil. Shortly, there will be conscious contact on a deeper level. This bridge of communication is now available because of the shifting you have facilitated on the Gameboard. This level of communication has not been available since the time that the Gameboard was first enacted. This new paradigm will be paving the way for the return of those carrying the crystal vibration. The next stage of evolution is at hand. You have made it so. We are so deeply proud of the work you have done to make this possible. The pain and turmoil you experience while in biology does not go unnoticed. It has paved the way for humanity to evolve. Heaven on Earth is at hand. The Grand Game is about to be won on Earth as it is in Heaven.

It is with our deepest love that we ask you to treat each other with respect, nurture one another and play well together... *the Group*

Chapter 14

The Second Wave

A Return to Power

A Week to Re-member in Cancun, Mexico

On the solstice you can see the snake (shadow on side of the steps) going from the top of the pyramid into the Earth signaling that it was time to plant or harvest. This Mayan technology dates back 8000 years. Here you can see the snake over Steve's right shoulder.

The Second Wave

*J*n May of 1997 I received a strong nudge from the Group to start a web site as a place for all Lightworkers to connect. This was anything but a gentle nudge and they also gave me several suggestions. One of the things they told me was to put the words "Second Wave" on the front page somewhere. I did as they recommended, but I never really knew the meaning of the words. At the time, I assumed this was to label the web site as part of a second wave of teachings. Looking back, I can now see that this was rather arrogant on my part. After all, these are the same entities that would not let me call them anything more flamboyant than "the Group". Obviously, fancy labels do not impress these guys. So what **was** the "Second Wave" all about? Actually I am just beginning to unravel that now.

It seems that it was important to plant seeds when we first started the site, and this is why they asked me to place those unexplained words on the web page. This is a typical way that spirit introduces important information to Planet Earth. This has been done in the recent past with words like "Light" and "Angels". Even now, they can be seen planting seeds in our conversations with words like "Crystal". All in all, I thought it was really funny that no one ever inquired as to the meaning of the words "Second Wave" on the web page. I'm glad no one ever asked. I really hate looking foolish.

The Group:

We are deeply grateful for the opportunity to address this gathering of awakening masters. As the veil between our dimensions continues to thin, you are beginning to see glimpses of your magnificence and your power. This has been our highest objective, for the time is drawing near for you to fully claim your power and use it in the creation of your next reality. *The love and respect we have for you are not always describable in your words. You have selflessly agreed to play a Game full of confusion and dichotomies for the higher good of all. With this confusion you often feel the true pull of spirit within your own being, only to ask if it is your own imagination. We understand the difficulties you endure for the sake of the Game, and we tell you that you are celebrated and esteemed for playing your part so well. Many eyes are now upon you for the events in progress. Yours is a job well done, and even though you may not yet understand what we say, we now tell you that you have already won the Game.*

What we will discuss in our time together here is of great importance to the planet. The time is now right for the grounding of this information. The Second Wave of individual empowerment is now forming and will soon begin to sweep across the planet. We have told you prior of the shift in direction taken by humanity in 1945. This was the most decisive event that led to the opportunities now being presented. This was the first indication that humanity was choosing to move back into empowerment. The wave of energy encircling the planet will help you take the next step into evolution as spiritual beings. This wave will be available as a result of the work of those who are grounding the energy on the Gameboard. Many have added their vibration to make this possible, including the cetaceans.

For eons, the grounding of energy on the Gameboard was accomplished largely by the custodians you call dolphins and whales. These cetaceans are descendants of your parental races, and they have done an excellent job of holding the energy of the planet during the Game. As humans began to awaken to their powers this task was joyfully turned over. It was always your contract to hold this energy, and there was great celebration as this torch began to be passed. During the last few years many have been awakening to take their power and lend their assistance in this process. The changes you have undertaken to enable you to hold this energy have caused many of the difficulties and hardships you have been experiencing. Now, the corner is about to be turned, and with a wave of the collective hand it is now possible to walk comfortably in your own empowerment. Thanks to all of you, this wave is building on the planet. If you add to this wave of energy, it will carry you easily into the next stage of vibration.

The Second Wave, as we have termed it, is about moving from a pattern of follow the leader into one of individual empowerment. The paradigm that you have set for yourselves has worked to some degree in the lower vibrations of your past. Now, as you have moved to a higher vibratory level as a collective, it is appropriate for you to begin a new paradigm that reflects your own individual empowerment. This is the reason that very few of your organizational structures still accomplish their intended purpose. Many of the systems intended to protect people have effectively separated them from their power. In giving way to these efforts, those you were attempting to protect, have lost touch with their own power of creation. We wish to re-mind you that you are in fact finite parts of the infinite Creator, and as such have all the power of the Creator. As you begin to re-member this power and express it, the old paradigms will begin to shift. This is the basis of what we have termed the Second Wave of Empowerment on the Gameboard of Free Choice. Now is the time to walk every step fully in your own power as you re-member your heritage.

CLEARING EMOTIONAL ISSUES TO CARRY THE LIGHT

Many Lightworkers on the Gameboard have been busy clearing their own emotional issues, thus allowing them to carry more light into the planet. This work is most highly honored, for it influences many more than you may be aware. The process of clearing your own energy to enable more light to travel is what we have termed *"Lightwork"*. *This emotional clearing will also enable you to move more easily into the Second Wave of Empowerment. Because of the work you have done in this area your sight is clear and many opportunities will be made available to you. One of these will be the opportunity to work with others choosing to make the transition to the higher vibrations. There will be many chances to help others adjust to the changing times.*

Change is always the precursor to improvement, yet it also brings discomfort to those clinging to the old ideals. As the Second Wave of Empowerment becomes rooted in the experience of planet Earth, there will be those who may experience difficulty, especially if they are unaccustomed to holding their power. The task of those we call Lightworkers will be to lead the way to advancement, first through their own experience, and then through assisting others. Yours will be the special task of lending a gentle loving hand to those struggling to step into their power. Some may see themselves as victims, as it appears their support is being pulled from them. They see that the crutches they lean against are all that keep them from falling. It is yours to show them in all empathy that the crutches they cling to so desperately actually keep them from walking. Helping each other to find the power within will be the focus of this Second Wave. Each person on the planet has a special gift to give. Find that gift, and they will find their power. Lead by example, and find first your own gifts, setting the energy for others to follow.

PERSONAL EMPOWERMENT - THE PATH TO LIGHT

Personal empowerment will in turn, focus global attention in new directions. As more of you step into your power, your organizations and governments will expand to make room for the empowered human. An ancient form of government will re-emerge and find its own balance naturally. This is the true form of government that once organized the great continent of Lemuria. In many ways it is actually a form of non-government. As with all truth, this ideal will find its own balance and flow naturally to all organizations that wish to make the shift. Many businesses have already begun to employ some of these principles with great success. They understand that to fully empower their employees will bring rewards to all. As an employee is allowed the freedom to move into their passion and empowerment they will produce benefits for everyone. This is most easily seen in the area of business because of the predisposition of these organizations to center their energy. Most of these organizations are formed with the clear intent of profit. Although this may appear selfish, it is this focus of centered energy that leads the way for its existence in the higher vibrations. Focused intent and clarification of motivation clear a path for manifestation to easily follow.

BUSINESS AND GOVERNMENT APPLICATIONS

Much will be forthcoming about the new organizational methods in the times ahead. Much will be re-membered from a variety of sources. We ask you to take none as the complete truth by itself. This is an area where many flavors of the truth will be necessary in order to glean the fullness of the information as it comes back onto the Gameboard. Private business will lead the way and governments will follow as the pressure builds to create an environment that enables individual empowerment. Prompted by advances in technology, this trend has already begun. As humans step into their power they will change the paradigms they have created. We will be dropping seeds of these organizational methods along the way, yet it is not for us to present this information in its entirety. It is important that each of you use your own powers to bring this information back into play. This is reflective of the Second Wave, in that information will come from the collective, rather than one source, thus making space for more individual empowerment. As always, *we will be here with gentle nudges as they are requested.*

The businesses of the world will soon provide a new standard for governments to follow. The shift toward a global economy has been underway for some time. The monetary system has accurately reflected the Free Choice aspects of the Gameboard. As a result, free enterprise systems reflect the natural flow of Universal energy. By emulating this Universal energy, a natural balance has begun on the planet through economics. It is with great humor that we inform you that because of this balance, a global war is no longer possible on the Gameboard. If this were attempted it would soon be discovered that one was attacking themselves. The fingers of economic interweaving have also set the stage for a global community to come forward. Global economy is setting the stage for true global community.

LEARNING TO LIVE TOGETHER WITH POWER

From an energy perspective, it is seen that all things are connected on the planet. This energy is accurately represented in global economies. As each individual re-members more of their own power and takes that power, the egos will become less engaged in government. With the ego less involved in this process, it will be a natural step to return to a system of government that not only makes space for empowered humans, but also makes space for other governments to co-exist peacefully, and even to nurture one another. Less will become more in government as humans begin to hold their own power. Humans will take responsibility for their own reality and refrain from leaning on their governments for their support. This has been a misdirection of energy that was propagated by governments in the lower vibrations of the planet. These actions do not support individual empowerment and must now shift.

This will take effort from each of you. All will begin to carry the new power and the responsibility that goes with that power for creating their own reality. This shift may be difficult for many, as the egos do not release easily. This can cause stress for those resisting the change. The ease of transition to new systems in government will depend on the resistance to inevitable change. We re-mind you of this important fact: _One is only able to fall if one is leaning._ Focus your energy within and find your own empowerment. Lean on your own inner empowerment first and foremost.

CENTERING YOUR ENERGY - FINDING YOUR PASSION

In your search for the paths to empowerment, let us nudge you to look in these areas. First, we ask you to find your passion and move into it fully. Finding this elusive passion confuses some, as many do not know where their passion lies. To begin this search it is helpful to see where you stand in your own heart. This confusion is often present because of a misdirection of energy. If one does not know their own position, even the best directions will not get them to their destination. We ask you to check your own position by observing the vantage point from which you view your world. Is your energy centered on yourself or are you viewing your world from the eyes of others?

Only by placing yourself first in all areas will you be able to give to others. We re-mind you that it is not possible to give from an empty cup. Think of this not as selfish, but as self-first. This is centering your own energy. Understand that even the most noble in your midst have had a "self first" motivation. In truth, even those who dedicate their lives to the service of others, do so, either because of the feelings they receive in return, or because they are clearing karma. When the energy is traced to its source it is seen that each one has a motivation that is self-first. There is no judgment about this. When your energy is centered, there is a natural flow and you are in harmony with the Universal energy. Be honest with your motivations and have the courage to treat yourself well. The God within you deserves the very best. Nourish that part of yourself and watch it flourish. Have the determination to find the crystals on your path and actively pursue the passion in your life.

Many on the planet often sacrifice themselves for the benefit of others, thinking that this is a selfless and noble act. In many cases, what they have really done is to teach by example that true happiness is not attainable. If your actions speak louder than words, are you "speaking" abundance, or a belief in lack? Are your actions "speaking" of holding your own power? Or, are you looking for things from the outside to be added unto you?

The natural flow of the Universe is the flow of energy seeking balance. Apply this search for balance in all actions and you will find yourself being assisted by the Universal energy. This is what you have termed "Being in the Flow." Balance the needs of those you are choosing to help with your own needs. When you do something for another, examine closely your own motives and observe the end result. Giving "freely" of yourself may actually have the effect of encouraging others to become reliant on your energy. This takes a situation of misdirected energy and amplifies it, making it worse, even though your actions were intended for their good.

EMULATING UNIVERSAL ENERGY

Emulate the Universe in the process of finding natural balance. Allow each person to achieve their true balance by feeling the support, or lack of it, from the Universe around them. Center your own energy and do not become so entwined in another's as to take it as your own. To do so is a disservice to that person, for it keeps them from seeing their true reflection. Approach this with compassion, for it will help you find <u>your</u> natural balance as well. Work together with the Universe and in place of making it right for others simply help them interpret the feedback from their actions.

The balance we speak of is the balance that brings the energy that is necessary to create your vision of Heaven on Earth. Only when you are centered enough to fully create your own idea of Heaven on Earth will it open the door for others to create their vision alongside your own. One unto another in this fashion, until the critical mass brings Home to your side of the veil in the re-creation of Heaven on Earth for all. Such action is the process you are calling the ascension. We call it "Winning the Game" and it is underway at this moment.

Look to those that have overcome great obstacles in their lives to see how it was done. They will tell you that it was simply a process of changing perception. Look at any situation in your lives through eyes that support a belief in lack, and you will see roadblocks. Look at that same set of circumstances through the eyes of one searching for possibilities, and you will create endless possibilities. Moving into these potentials is the beginning of your creation of Heaven on Earth.

FINDING THE CRYSTALS ON YOUR PATH

When you scripted the phase of the Game you are now playing, you set many crystals on your path to mark your way. The crystals you have placed in your path carry the vibrations of Home. When you find these crystals, they feel wonderful and re-mind you of the wonderful vibrations of Home. Many of you are good at locating these crystals, yet when you find them you often resist the message they contain. These crystals mark your path of least resistance. The part that is confusing for most of you is that these paths also contain a great deal of joy. Your judgments often tell you that it is not possible to be on your path and experience joy in this manner. The ideals you carry have told you that you must work hard and sacrifice in order to achieve success. We tell you that you are very good planners and that your path is not that difficult. Look for the possibilities *that contain excitement and passion. These are the markers that you have set up for yourself. Moving into these areas of passion and joy is what will lead you to your highest potential. Being in your highest potential will also attract your greatest abundance. These are the tools for the creation of Heaven on Earth. When you have the courage to center your energy on yourself and move into the areas of your greatest joy you will quickly win the Game.*

You have made it so and we are so very proud of you.

It is with great honor that we ask you to treat each other with respect, nurture one another and play well together... the Group

Phyllis Brooks

Chapter 15

Time

Does anybody really know what time it is?

Barbara & Steve in the Tulips of Holland

Time

Lightworkers everywhere are experiencing difficulties with time. They are finding it especially difficult to get things done within a "normal" timeframe. For many, it seems as though everything is taking a little longer. In addition, there is an anxiety in the air that gives us the impression that we must hurry and get on with it. It leaves us with the underlying feeling that if we don't hurry up and move into our contracts the ascension will happen without us.

The Group addresses this by saying that the feeling of anxiety is a biological side effect of our increased perceptions. Here, they offer us insights to Universal time and how we can begin to experience it.

The Group:

We are again deeply honored to address this group of masters with the title of Lightworkers. This is a chosen title, and is bestowed upon all those requesting to carry it. It reflects a willingness to consciously carry light to planet Earth, and this attitude is the only prerequisite to holding this title. As you know, we are reluctant to use titles of any kind. We find humor in the fact that humans place so much importance on titles. We see that oftentimes humans accept messages of questionable integrity because of an elaborate title attached to it. We ask you to give the attention to the love content of the message and use your own discernment with all information that enters your field. This is the greatest expression of the God within each of you and will help you maintain your balance in these times. We tell you that even with this aversion to titles, we are honored to call <u>ourselves</u> Lightworkers. Instead of using this title to place us on some imaginary ladder of importance, we use it as an affirmation of our intent. We invite you all to do the same.

We are aware that often you feel distress and discomfort as you stretch your beings to incorporate the higher vibrations. We tell you that you are loved beyond your understanding for this act. At this juncture, we wish to point out that the vibrations of those we are addressing have changed considerably since the time we began this flow of information. The vibrational advancement accomplished by this action paves the way for many others to follow easily. Your willingness to move forward and take the lead for the benefit of the planet is what endears you so to us. We are proud to be here offering this information for your individual discernment. We are here only because you have asked. In asking, you have exercised your power and also have allowed us to fulfill our contracts on a very high level. We humbly thank you for this opportunity. We are deeply honored to be a part of the Game as humanity steps into the next stage of evolution.

PERCEPTIONS OF TIME CHANGING

In this session, we wish to speak of your changing perception of time. On the Gameboard, the mechanics of polarity color your perception. We ask you to understand that this was a necessary part of the Gameboard. Looking though eyes tainted with polarity, you can only see the dimension of time known to you as linear. Your existence on the Gameboard has been one of being finite pieces of the infinite whole. Since it was not even possible for you to understand the concept of "infinite" in your current condition, it was likewise not possible for you to understand the perception of the Universal "<u>now</u>." This is changing as a result of your forward movement. Much is now available to you that you are only just beginning to be aware of. This is a time of return to your power, and as it floods back into your consciousness many doors open. One of these doors is a new relationship to what you call time.

WHY YOUR BIOLOGY IS SHIFTING

Those of you who have chosen to step fully into plan B have set into motion a shift deep within your own biology. Through your intent to move forward you have begun the process of shifting your biology at the cellular level that you know to be the DNA. We have spoken of this in prior messages, and told you that this was actually a return to a state that you once knew when you first formed the Grand Game. Returning to this state of higher vibrational biology is bringing about changes that are just now beginning to show themselves.

As your biology settles into the higher vibratory levels, you may experience heightened states of awareness and sensitivity to energy. With each new level of vibration attained on the planet, a short period may be needed for your biology to adjust. This is a normal process and is not to be feared. It is simply your biology adjusting. These symptoms will often show as lethargy, or simply a lack of energy. Other times it will be body aches and energy sensitivity in many areas. This can be termed the **vibrational flu**, because the symptoms are similar. In addition, your heightened sensitivity to energy will often show in your emotional bodies. During these vibrational shifts it will be commonplace for Lightworkers to experience heightened emotional states.

Please be aware that these are not only a direct result of the shifting of the planet, but also the shifting within your own biology. The emotional body and the physical body are so closely linked that subtle changes in the DNA often create this heightened emotional state. These two causes are creating difficulties for those of you leading the way to vibrational advancement. We understand your pain and ask you to be patient, as this will ease quickly. Each time you experience these shifts it will get easier. With this physical and emotional challenge there also comes a strengthening that will make it easier the next time. For now, please view it as a growth process, and know that it will not last. Grounding and breathing techniques will help balance this adjustment process.

As your biology shifts to higher states your true senses become more attuned. In reality, you are moving from a dense body to an ethereal body. In doing so, you are also moving from a field of linear time to a field of Universal, or "now" time. In your limited perception you are only aware of a very narrow band of vibration. Most animals on the Gameboard have a sense of perception that extends well beyond human capabilities. They can see, hear and smell far beyond your ranges of perception. As your own DNA begins to realign and shift to the higher states, your range of perception steadily increases. Most of these increases will not be felt in traditional ways. It is not as if your body will hurt as a result of shifting DNA. Instead, this shifting will begin to show itself in more subtle forms. One of these forms is in becoming aware of other dimensions of time.

A JOURNEY THROUGH THE HALLWAYS OF TIME

Let us lead you on a journey to illustrate the reality of Universal time in relation to your linear time. Being limited to three dimensions, you see time as linear. Visualize yourself traveling very fast through a long hallway. Even though you do not see the end, you have awareness that there is a beginning and an end to this hallway. This illustrates your being in finite form and reflects the polarity of the Gameboard. You have been traveling down this hallway for the duration of the Game thus far, and you are quite familiar with the passageway. Advancing down this long hallway, you make a conscious decision to raise your vibrational level, and in doing so, you start to increase your range of perception. Now, in the hallway, you look to your right and notice that there is another hallway going off at a ninety-degree angle. Even though it was a very quick glimpse, you are sure of what you saw. You quickly search your memory banks for an explanation, with no results. Lacking sufficient information, your ego intercedes to inform you that it was just your imagination. Temporarily accepting this explanation, you return to the journey down the hall of linear time.

Before long, you catch a glimpse of another hall, this time off to your left. Further down the path you see another one going off directly above you and then one below. This time, it is not possible to dismiss this as simply a result of your imagination. Make a mental note that all of these passageways are at right angles to each other. You wonder what is down these hallways but are unable to explore any further. Your speed down the hall continues at the same pace as before, but your range of perception seems to be increasing. You can see more than you have ever seen before. Now, as you pass these other hallways, you can see that they are very similar to yours. These are actually other dimensions crossing each other at right angles.

MORE ON CIRCULAR TIME

Now, we pull you back into deep space to view this collection of hallways. Moving back, you can see that each one of these hallways is slightly curved. This curve is so infinitesimal that even from inside the curve is undetectable. Moving further back, you can see that each of these curving hallways connects to form a very large ball. This is an example of circular time and also relates to the time-space continuum previously introduced to the planet through the Lightworker known as Einstein. Each of these hallways is connected to each other in several places as their paths naturally intersect. Even though you may see yourself as traveling down a hallway with a beginning and an end, this is not the case at all. With this new perspective you can see that it is possible to be anywhere, at any time, as long as you know which turns to take. With this information **time warping** is now entirely possible.

You are evolving at a phenomenal rate, and as you continue to *advance you are increasing your range of perceptions. With your expanded sensory range you are even now beginning to see down the hallways as you pass them. These most often present themselves to you as vague shadow figures out of the corner of your eyes. Perhaps you feel like there is someone else in the room, only to turn and see nothing. Your peripheral vision will be the first to perceive*

Universal Time

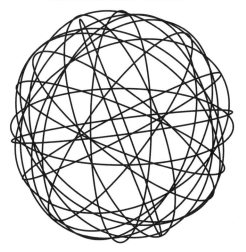

**Intersecting Hallways of Time where Linear time
becomes Circular, or "Now" time.**

The Hallways of Linear Time

With the expanded senses of the new biology,
humans are becoming aware of other dimensions of time
intersecting with their own. With these expanded senses
TIME WARPING is now a reality.

these other dimensions of time, because the veil is most transparent in this area. Please understand that this is not new. There have always been many dimensions occupying the same space. What is new is that your range of perception is now allowing you to perceive them. Upon closer examination, you soon find that not all of these other dimensions are equal to your own vibration. In fact, some of them are of a much lower vibration. Choosing to give your power to entities in these dimensions, simply because they are unexplained, is not in your highest good. For this reason, we ask you to apply the principles of discernment to all information you receive. Look for the love and trust first in your own knowingness.

THE EFFECTS OF EXTENDING YOUR PERCEPTIONS OF TIME

There is a natural biological reaction to extending your range of perception. The ego is still a part of your existence on the Gameboard, and as such you must balance it often. With these new visions and ideas entering your field, the ego produces a natural reaction to these new events. The result is that most leading the way in this advancement are now experiencing distortions related to their perception of time. One way that this manifests is an anxiety that leaves you with the feeling that you must hurry and move into your full power. This feeling gives you the impression that if you do not rush, the world will move forward without you. Some people feel this so intensely that it easily distracts their energy, demanding all of their focus. Know that this is a natural biological reaction; a side effect related to your new perception of time. You have enough time to do everything you came here to do. The fact that you are feeling this anxiety indicates that you are actually on track with your progress.

DISTORTING TIME

Another side effect is that your relationship to time becomes distorted. This often makes it difficult to accomplish tasks in a predetermined timeframe. Repetitive tasks will often seem to take longer than they ever have before. What this illustrates is that you are beginning to reclaim your power to control time. The problem is that most of you do not know how to exercise this power, and by leaving it uncontrolled it slips easily away, leaving you with a shortage of time.

WARPING TIME

Warping time in the higher dimensions is a matter of exercising conscious control. We will give you some of the basics and plant important seeds in fertile ground. We have told you many times that you are much more powerful than you see. So often you look to the outside for your power, only to find it waiting patiently within your own being. The many times we have spoken that re-membering your powers are most easily accomplished by adjusting your perceptions very slightly. Such is the case with time shifting. The biggest challenge to shifting time is to set your initial goals small enough to allow them safely past your ego and your own belief systems. At present, your time can be bent only slightly, as your collective belief systems do not yet support complete time travel. Therefore, we ask you not to attempt to manipulate time in a grandiose fashion. Start with regaining control over the daily tasks you know to be well within your reach. As you become more accustomed to intentionally controlling time it will become easier for you to gradually increase your objective. Set your goals small enough to be believable. Find something for which you choose to create more time. It is also possible to set your intent to get more done in a smaller amount of time. Stretching and contracting time are activities that will be used in the higher dimensions. Using them now on even a limited scale will prepare you for what is ahead.

We understand your attraction to the mystical. This is a direct result of the veil on the Gameboard as it keeps you from seeing your own power. If you wish to make this into a mystical experience please feel free to do so. All we ask is that you keep in mind that we are simply leading you back to your natural powers as finite parts of the infinite Creator. In the very near future these same powers that appear so mystical today will become commonplace and mundane. Make space for these, as you would brushing your teeth in the morning, and see how they can add to your experience as the Gameboard shifts.

How to Warp Time

We will now suggest a starting point. It is important to incorporate your whole being into the process. Before you begin this exercise, express your intent aloud to noticeably warp time during this experiment. Note the time as you begin, and then release all concerns about time and your relationship to it. A most effective means of uniting spirit and biology through intent is the use of ceremony. We will plant some seeds here for you to form your own rituals, as those are the most powerful rituals you can use. First, we ask you to center your thoughts. With your eyes closed, raise your eyes above the horizon, as if you were looking up while keeping your head level. Do this for as long as it feels comfortable, and do not be overly concerned with holding this position. If you feel your concentration slipping away during this exercise, simply raise your eyes and start again.

Focus your thoughts within and allow the outside world to slip away. Please re-member that it is not your choice what thoughts enter your head, as you are simply tapping into the flow of Universal thought. Even though you may not be in control of what enters your head, you are in control of what stays there. It is always yours to decide which thoughts reside in your consciousness. This idea of conscious thought selection is the basis of reclaiming your true power of creation. As these thoughts enter your head, simply acknowledge them and make a conscious choice to let them flow out the other side without attachment. Do this for a time to become accustomed to the feeling of unattached flow.

Now, we will ask you to imagine a color for each thought as it enters your head. This is not a judgment, but simply an impression. As each new thought enters your head, notice its "color" as it moves through. Release the need to judge the colors, for there are no good or bad colors. Now, take a moment to choose only the bright colors. As these flow though your consciousness, allow the brighter colors to stay for a time and visit with you as you move the subtle colors through. Now, choose to release all colors. Now, as the flow continues choose only the subtle colors to reside within you. See each thought of subtle color as it passes through. Now allow only those thoughts that represent the many shades of Blue. Then allow only the thoughts of the Greens, Reds and, finally, Yellows.

Now direct your thoughts to the use of time. As this thought enters your head become aware of its color. Now take this thought about time and intentionally change its color and shift the hues that you see before you. Now you are seeing your perception of time change. Take this thought and move the color to one that suits you. Make it your special color. Notice any sounds or smells that may accompany this thought about time. If they are not to your liking, alter them in a way that makes them pleasant for you. Take a mental picture of this thought and remember the color, the smell and sensations surrounding this thing called time. Make a mental note to pull up this image at any point in the future when you wish to alter time. When you are finished, come back and breathe in the experience. As you release the breath, know that you have just discovered a tool for your higher good. Use it often, and you will become well practiced and able to alter time more efficiently. Note the time as you return and see how long you were gone.

This is but one way to begin taking charge of time as you have known it. There will be many more techniques surfacing as you begin to play with this. We ask you to openly share what you learn and help each other to understand the possibilities. Each one of you carries a special piece of the puzzle. The whole picture can only be seen by sharing these pieces.

You are at a stage of your development when it will be helpful to re-member that the spirit within is with you in everything you do. Connecting spirit and biology through the use of ritual will help you live consciously every moment of every day, walking in unison with your higher spirit. This is the re-membering of your power and the beginning of the marriage to your own higher self. This re-union is what you have been calling the ascension. This process has begun.

THE MASS AWAKENING

The masters on the Gameboard are awakening in force. It is exciting for us to stand on the sidelines and offer assistance. The objective of this Grand Game was to see if you could re-member your powers and use them to create Home on your side of the veil. You cannot imagine the excitement and honor we feel that this Game is now being won. The love we have for you is not explainable in words, but can be described only as a spiritual family connection. We are integral parts of one another. The feelings that we so often send you in these writings are the only way we can communicate this love. Please accept these feelings, and know that as they resonate deeply within your own heart they are, in fact, crystals that you have placed on your own path to re-mind you of the way Home. Please know that we are always here for you with love and *information as you request it. Even though you often feel alone on your path, we are at your side every moment of every day. Ask, and we will come to you. We will come to you in any form that you will accept. All it takes is your intent and your willingness. You will know us by the family connection and this feeling of deep love within as it resonates within your own heart. We proudly stand at your side as you create Heaven on Earth.*

It is with deep love that we ask you to treat each other with respect, empower one another, and play well together... the Group

It's a great time for the planet. It's a great time to be here!

I am honored to be here and to claim for myself the title of Lightworker.

Chapter 16

Biology

The Evolution of Humanity

Biology

Each month I sit and write these words. Sometimes I write in my office at home and other times while riding in a car or on an airplane. Wherever it is, I treasure the time I spend in this activity. This is very invigorating and I am aware that when I am in this state my body stops aging. There are times when it is very difficult for me to sleep after channeling. I am also aware that I often receive healing during these times. The same is true for the live channels I do at the seminars.

These messages are not about the Group, or me, rather, they are about finding your own empowerment. Me? I'm merely an ex-contractor who was tapped on the shoulder and asked to carry a message. It's not a complicated message, but it is grand in scope. The Group has told me that the contract between us was made a very long time ago. They also said that much of what I have gone through in my life thus far was to lay the groundwork for what I am now doing. When they said that I breathed a sigh of relief. I didn't think I would make it through at times. I know my mother shared those same thoughts. The funny thing is, if you had told me a few years ago that I would be traveling all over the world channeling messages from entities on the other side, I would have laughed at you. Now, here I sit on an airplane on the way to Vienna to channel before a gathering at the United Nations. The interesting thing is that when I started channeling I had this feeling in the back of my head that I had been waiting all my life for this.

In putting out this message the Group emphasizes the importance of connecting spiritual family. This is the idea behind the web sites and the seminars. These are ways to bring family together. When we hold hands and connect we begin to see who we really are. This is especially important as we begin the many changes that we've asked for. It's one thing to ask for the changes to lead us into ascension. It's another to walk through them as they begin to unfold. In these difficult times it is helpful to connect with each other and support one another. The physical changes to our biology are something that we all share.

The Group:

Greetings from Home.

We will now begin all of our communications in this manner, for it reflects our true connection to each of you. As the veil continues to thin you are coming closer to understanding your true origins. This greeting illustrates the cosmic wink that helps you re-member Home. The love and honor we feel for you are beyond the words you have devised for your Gameboard. It is our honor to connect each of you to the pure vibrations from Home and help you to re-member your own heritage. It is our sincere hope that as we provide this connection you will re-member more of your own power and use it in your daily lives.

The evolution of humanity on the Gameboard of Free Choice is honored more than you can know. The effects of the actions each of you take at this time will resonate throughout the Universe. Already, you have brought great honor to this family of light. You are the architects of light. We thank you for this opportunity to once again hold a mirror to you to help you re-member your magnificence and your true nature.

The game as you originally scripted it has shifted levels and the new game has now begun. Many have felt the heaviness on the planet at this time. Many of you have begun the process of evolution through your DNA/RNA. Therefore, you have an increased sensitivity to these energy shifts. As each one advances their own vibrational levels, the vibrations of the planet rise in a collective harmony. This is achieved in increments. Prior to each new level, or increment, there is a compression of the energies that produces the heaviness you feel. With the shifting of your DNA you are very sensitive to the emotional energy of the planet and this is what you are feeling.

FEELINGS OF SADNESS ON EARTH

The Earth is an integral part of yourself, and these connections go far beyond what your senses can discern. The Earth is also moving into her next stage of evolution. Just prior to achieving a new vibrational level there is a momentary sadness as the final steps are being made. This would be equivalent to your moving into a new home. Your focus and excitement about your new home usually take most of your attention until the time you actually leave your old home. In that final moment you must say goodbye, and a bittersweet sadness pervades even as you walk into your new home. Much the same as this illustration, we ask you to feel the sadness and move forward into your new surroundings. If there are times when it seems to be overwhelming, reach out and take the hand next to you, for there are many moving into new homes at this time. To fully comprehend the new light, a period of darkness must precede the experience. Holding hands in the darkness will enable you comfortably to usher in the new light.

Learning the new rules of this game and the mechanics represent difficulties for many. Each of you has chosen a path of change. We see this and stand at your side when asked to help you walk through these changes. It is the act of shifting this energy that you see as struggle in your life. It is this same struggle that we see and label as Lightwork. This is the work that will enable each of you to walk every step in unison with the true power of your higher self. At every moment that you purposely create your reality, you pay tribute to that part of the Creator that is within each of you.

As more of these increments are reached, it will become easier for each of you to walk through these times of change. For many, these periods of darkness will eventually provide opportunities for the expression of your own light.   If you find yourself unable to separate yourself from these emotions, then balance yourself by offering your hand to another.   As you extend your hand to help them balance, also will you then find yourself anchored and balanced as well. Offering your hands to others will be a valuable service during these times. When you feel you are empty inside and have nothing of value, these are the times that we ask you to reach out and offer your assistance to others.   It is this honoring of the connection to all other things that will allow you to see your own gifts more clearly in this grand new light.

FEELING STUCK

Many feel they have been left standing with no support as the new paradigms move into place. We ask those people to be aware of the perspective from which they have chosen to view this enigma. You have stood and gazed into a mirror and seen your own reflection for a long time. With the vibrational advancements the mirror has moved. Now, as you look into this mirror you no longer see the same reflection. We ask you to reach out to others within your field, as they will help you to adjust the point of perception from which you view the mirror. Above all, this is a time for re-membering your original spiritual family.

The honoring we have for you is not describable in your languages. You look at yourself and think this is about you. Your egos have made you believe that this is limited to a personal experience. This is an illusion based in polarity on the Gameboard. You look into the mirror with the critical mind of your ego and the Universe answers your thoughts with a resounding **"and so it is."** We look into that same mirror and see your reflection as a very special part of the Creator that is held only within you. We look into that same reflection and see the very best parts of ourselves. We know it is difficult and confusing for you. We see the dichotomies that you must choose between. We see the conflicts and illusions of polarity that face you every day. We see the loneliness you feel as you draw even closer to Home. We see your struggles and we love you for them. It is through your handling of these challenges that advances are made, reaching far beyond your own dimensions.

THE PENDULUM OF HUMAN ADVANCEMENT

The changes you are now making will set the stage for the evolution of humanity. We often use the illustration of the pendulum with the movable fulcrum to show the advancement procedure. The pendulum is pulled to one side to set the experience into motion. Each of you adds to the momentum by adding force to each swing until it reaches a critical point. At this point the fulcrum, or pivotal point of the pendulum, moves to the next notch. This movement to the next notch of the fulcrum is what you have called evolution. When the fulcrum advances in this way it cancels the swinging movement of the pendulum and so a new starting point begins. From an historical perspective, these evolutionary shifts have happened in the blink of an eye. What is often forgotten is that many dramatic arcs of the pendulum were needed to produce these shifts. This is true of all evolution and is not confined solely to your planet. Now, you on the Gameboard of Free Choice are once again at a stage where the fulcrum is about to move to the next position. We will focus this session on some of the immediate biological changes that face you as you move into this next stage of your evolution.

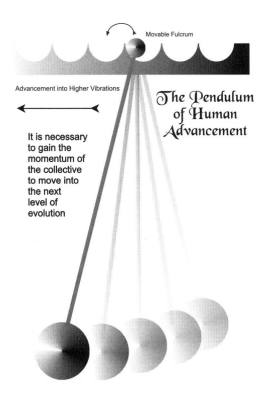

Movable Fulcrum

The Pendulum of Human Advancement

Advancement into Higher Vibrations

It is necessary to gain the momentum of the collective to move into the next level of evolution

THE EARTH CONNECTION

The levels of the game are changing. We will address the changes that lie before you, but it is helpful to understand that the entire human collective will experience these changes on some level. This includes the part of you referred to as Gaia, or Mother Earth. The connection that binds you together with the Mother extends far beyond your understanding. It is not that you are limited to be forever on the planet Earth. For it will soon be yours to travel far beyond the physical limits of the planet. Still, the energetic connection remains constant. It is this energetic connection that causes you to evolve together as one. The advancements you make in raising your own vibration have averted many of the Earth changes that were originally scripted.   Similarly, this advancement has lessened the severity of your own physical changes.   Still, there are many alterations that will be necessary for you to carry the higher vibrations of Light while in physical form.

The evolution of the game is represented differently as perceived by the many parts of the whole. From an energetic perspective, you are moving from a field of polarity into a field of unity. From a dimensional level, which is related to what you call time, your first move will be from the third level to the fifth. From an Earth perspective, you will be re-arranging the Gameboard to accommodate the new game. This will result in both dimensional changes to the vibrations of the Earth as well as vibrational changes to the dimensions. Let us recap this by saying that the shifts are occurring simultaneously on many levels. The Gameboard of planet Earth will very quickly move to a new vibrational level as it completes the shift to the next dimensional level.   The need for cataclysmic Earth changes has been transmuted, yet the Earth will still be raising Her vibrational rate.   She will simply make these shifts in a much less dramatic fashion. As it stands at this moment, there is a need for the Earth to relieve stress.   This will be accomplished by the Earth changes that you call earthquakes. You will determine the severity of these shifts on the Gameboard. These shifts will still be needed as the Earth shifts from a dense state to a lighter one. It is accurate to say that the Mother is moving from a physical dense body to an ethereal Lightbody.

Physically, it will be seen that the human form will change substantially as you move to higher vibrational levels. In this higher state of vibration the possibilities are beyond your dreams. This is the representation of Light that has captured your attention on your Gameboard. Your attraction to the word Light has set the stage for you to experience these physical changes.

EXPANSION, CONTRACTION AND THE BIG BANG

The beginning of the Universe, as you know it, was what you have called the Big Bang. Prior to this event, everything was in motion toward density. This event caused a reversal and began the process of expansion that has continued to this time. This represents the natural movement of energy within the Universe from dense form to lighter form. This is the evolution throughout all of nature that will be clearly visible on planet Earth in the near future. This is the action of the Universal energy seeking balance. In the illustration of the pendulum, all was pulled very far into density so that it could begin the movement into expansion. As the pendulum continues now to be pulled into expansion, it will soon reach the point where it will jump to the next level and the fulcrum will find a new balance point. It is important to note here that in both directions there was always Universal movement. First, the movement was toward density, then toward expansion. It is not by coincidence that the densest forms yet known to your scientists are called black holes, characterized by darkness. Likewise, it should not surprise you to see movement toward expansion equated to the word "Light".

On a physical level, your biology is in the process of rearranging. At this next turning point, your bodies will lose some of the parts that you have outgrown as new features develop. Most of you have noticed that as the vibrational levels rise, your sensitivity to energy rises also. During this transition this may present difficulties. After humanity has become familiar with the use of this information it will become one of your greatest strengths. You have senses that you use everyday of which you are still not aware. The sense of absorption is one that you use to absorb energy into your bodies. The closest sense in application is your nose and tongue as they absorb energy in a similar fashion.

HISTORY OF THE EARTH CONNECTION

When you first began the Game of Free Choice, you took form as ethereal beings. This Lightbody form allowed free movement through a wide vibrational range. As the Earth cooled it gained density. Humans in ethereal bodies found it difficult to stay grounded. Mother Earth responded by offering adjustments to your biology that enabled an energetic connection to the Earth. This was accomplished through assimilating the energy of the Earth through what you know to be eating. You began this process first by ingesting leaves and fruits, soon graduating to nuts and vegetables. As your new bodies gained density, the energetic connection between your biology and your spirituality became labored. Now that you are moving back into the higher vibrations it is being restored.

Now you have reached the portion of the Game when your biology in its current form will soon begin to change the energetic connection to the planet that is your biological mother. This time, the Mother is raising her vibrations through a process that you have known to be global warming. It was necessary for the Earth to raise her vibrations to begin her shift to the higher vibrational levels. This process is now well underway, and as it unfolds your biology will find it more difficult to connect in the ways to which you are accustomed. This is the juncture you are currently at in your evolution. Since the Earth is now moving to a state of higher vibrations and lighter density, your energetic connection is beginning to falter, and this is causing many of you stress.

CHOOSING LIGHTBODY

All forms of biology are products of Mother Earth. The Earth is a living, sentient being that is connected so close to you that you often do not notice the connection. The Mother is now at a stage of offering further advances in biology that will enable humans to carry more light than ever thought possible. This is the transition back into your original human form; to that which you have called Lightbody.

The choice is yours, and there is no judgment as to your choices. Most of you resonating with this information have already made the choice to step firmly into Plan B. There are many on the planet at this time that have chosen to step into Plan B that will also be leaving. There is much to do in Plan B that can only be done from the other side. The Children of Crystal Vibration will be entering through the reincarnation process and the first of these are now in preparation. The Crystal Children will carry these seeds of new humanity onto the Gameboard. Those entering with these seeds of advanced biology will have an easy time of moving into Lightbody. Eventually all humans will be born directly into Lightbody. Those wishing to stay will also have the opportunity to move into various forms of Lightbody, yet the path may be more challenging and will take a focused effort, for this is not an easy path. The transition will require the clearing of both the emotional body and the physical body. Once cleared, the Earth will facilitate the reworking of the DNA to set the move into Lightbody into motion. For many of you this process has already begun. To set this into motion takes only the stated intent to walk into Plan B and write your next script.

EASING THE STRAIN

You have averted many of the cataclysmic Earth changes that were to come through your inner advancement. Now we tell you that the same is possible for your own biology. You have set into motion evolutionary steps that will lead you to a higher existence. The first ones to take these steps experience the majority of pain. Your willingness to be at the cutting edge of this advancement and take these first steps places you among the elite of the healers. This act alone brings you honor and respect that is not possible to fully convey at this time. The most important tool we can give you at this stage is to tell you to stay close to those around you. This step into evolution cannot be taken alone. An important part of this step is the strengthening of the connection between all things. Separate yourself and the flow will be blocked. Allow yourselves to be vulnerable and to connect with others making similar steps. Offer your own experiences to others preparing to step where you have been. Make yourself available as a human in experience of the Grand Game. Allow room for mistakes in yourselves and others. We have likened your experience to being in a dark room, all looking for the way Home again. Understand that at these times of evolution some might get scared, falling into judgment, ego or fear. Some that have always led might feel themselves falling behind, searching for a new expression. The ego may take this as a threat and react with feelings of anxiety. Do not let this discourage you. Your job is not over, it is simply evolving and finding new expression.

As we have said before, at this time there will be many that feel the pull of Home and choose that path. For you, we honor your decisions, yet ask you not to rush into action here. Allow your higher self to orchestrate all, as many of you will find new contracts emerging in unexpected places.

WHAT'S HAPPENING TO MY NOSE?

Your biology is known to operate on a chemical electrical basis. We tell you that the nose has always been a source of electricity that enables this system to operate. Your biology balances the electrical energy naturally. The nose is constantly changing nostril dominance and this is the basis for the biological balance from an electrical standpoint. As you inhale through your nose, the air rushes past the hairs and mounds of flesh, generating an electrical current. This energy is then stored on the opposite side of your spine. Breathing in the right nostril places a charge on the left side of your spine, and

vice versa. These electrical supplies feed the brain on the corre-
sponding side where the charge is stored. It is therefore possible to
feed one side of your brain energy in anticipation of using that side.
This information has been on your Gameboard for some time, yet
few use it to their advantage. This information will be even more
useful in the higher vibrations because of the increased sensitivity to
energy in all areas.

The nose is also a particle receptor, much like the tongue. Now,
as humanity begins to move even further into the higher vibrations,
you will find the uses of this receptor expanding. Your movement
into the higher vibrational levels has brought about a new sensitivity
to energy that previously has not been discernible. The life force
energy that is present in all things has very little validity in your current game. This is the form of energy to which you are now becoming very sensitive. The characteristics of this form of energy are
closest to emotional energy, and this is one reason you are so sensitive to the emotions of others. The particle receptors within your
nose are becoming more effective as they are now able to absorb
much smaller particles. Many have noticed changes within their
nose and a heightened sensitivity to smell. The sense of smell
becoming very acute or the constant running of the nose at this time
indicates movement in this area of vibrational advancement.

The sensitivity of your nose is growing beyond your understanding, as this is also now used to absorb pranic energy. Much will
be learned from this as humanity moves forward and steps into the
next stage of evolution. This is the balance of a new system of
chakras that are developing in the human form as it evolves. The
sense of absorption with your nose is the balance to the chakra that
is forming at the base of the skull in the back of the head. At present your mouths and noses are both used for absorbing oxygen.
They are also close in proximity and connect using the same general
passageways. Such is also the case with the new energy receptors.

It will soon be commonplace to store much larger amounts of life force energy within the biology due to these receptors. Those leading the way vibrationally will use their noses to discern many types of energy. This has been seen thus far only in the most obvious types of emotional energy. It has been possible in your current state to smell excitement or fear. There are animals on the Gameboard that have used this evolved sense of energy discernment for some time. From your perception it seems as if they can actually smell danger. This is a form of energy absorption to which you will soon evolve. From our perspective, it is known as the Breath of God. Being able to hold larger amounts of the Universal Life force energy will set the stage for comfortably holding much higher vibrations within your biology. These are the next steps in human evolution and they are in motion now.

EMOTIONAL SENSITIVITY

Vibrational sensitivity will also show in many other areas. Emotional energy is one of the strongest forms of this subtle energy. Your sensitivity to energy has been increasing rapidly at this time. This shows very clearly in your sensitivity to emotional energy. It is one of the more coarse forms of this energy, and therefore is the most easily discernable to humans in the lower vibrational forms. Your sensitivity to this energy will be one of your greatest assets and will be a tool that you will use often as healers. As this sensitivity increases there will be times of adjustment that may be difficult for some. As this energy enters your newly forming receptors it is in a clearer form than you are used to receiving. As a result, the recipient is often not fully able to discern what energy belongs to whom. This is a temporary condition that will soon pass as one learns to assimilate this information. Proper grounding of the biology is the most effective way to balance this condition.

Help! Why am I Gaining Weight?

The effects of moving to higher vibrational levels on the biology are not the only challenges you are experiencing in your biology at this time. Your connection to the Earth is getting stronger during this advancement. The Earth Herself is moving into new vibrational levels and those changes also have a direct effect on your biology. These advances with the Gaia are directly related to your collective vibrational advances. These shifts occur in increments. That is what causes the difficulty with the reaction in the human biology. When the advances occur, the higher vibrational beings often feel the surge of energy within their own shifting biology. They are often sensitive to the energy because of their own advancement. The biology feels this assault on subtle levels and reacts to protect itself in varied ways. One way the biology protects itself is to coat the cells with fat for protection.

During the shifts these increments have been steadily increasing. As a direct result, much of this weight you have gained for protection will not completely fall away between these shifts. This is a temporary situation and will begin to alter as the collective vibration of the planet continues to increase. A balanced diet and exercise can minimize the challenges during these times. Balance in all things is important, but we caution you that the base of your power is held within your own thought process. Fear the changes and they will become an increasing challenge. Worry about bodily changes and you will create the most difficult scenarios. The veil is thinning at a rate never before seen on the Gameboard. Now is a time to pay particular attention to the thoughts that you allow to reside within. The battle cry of Lightworkers at this time is the same as it has been through the ages. It is simply: FEAR NOT. Set your tone and then walk purposefully into your future.

CHANGES IN YOUR BIOLOGY

As the body adjusts to the higher vibratory conditions it begins changing to accommodate the new environment. Some of these other changes can cause alarm if one is not forewarned. Most common are sleep changes and waking in the night, heart palpitations, shaking or episodes of vibrating, temporary, irregular heartbeats, hot flashes not attributable to menopause, shortness of breath, short-term memory loss, and early menopause. The geographical location in which one resides will have a bearing on which of these symptoms may be prevalent. This has to do with the movement of the magnetic grid lines as the Earth proceeds with her advancements. These movements are needed to support the higher biology to which you are evolving. Your strong connections to the Earth will now reach new levels as both of you move into higher vibrational states. There are healers on the planet now that will be able to point you in the right direction of finding these modalities and choosing ones that work best for you. Seek them out and follow your heart. Be responsible for your own healing; yet take what is offered when it resonates with you.

VIBRATIONAL HEALING

The symptoms of vibrational advancement will open the door for many areas of vibrational healing to take hold in the consciousness on the Gameboard. Vibrational healing will emerge and help to balance these symptoms. Let us encourage you to seek these modalities and use them now to balance. New biology will bring even more modalities and expressions of vibrational healing. Currently, these are the expressions of light, sound, aroma, absorption, and magnetics. What we refer to as absorption is the method by which you receive pranic energy. Much more will be unveiled in the near future about these and other expressions of vibrational healing.

Let us now validate that there is a vibrational signature to each ailment or disease that can be held within your biology. Introducing an harmonic vibration can eradicate the disease, because it is not possible for two vibrations to occupy the same time and space. There have been many that have embraced this approach, yet none have incorporated the overall body signature into the equation. Each body has an overall vibrational signature, which is a compilation of all of the individual parts. By learning to read and then to alter this signature, one will be able to make safe and simple alterations to the biology. This is one of the areas in which you will see many advances in the coming times, because it is a natural way of assisting biology.

Healers of the future will work on a different basis than those in place at this time. The first and foremost difference is that they will hold their patients responsible for their own healing. Secondly, it will be possible to offer remedies with no side effects. Vibrational remedies will either be effective or inert. The exceptions to this rule are the remedies that are used over long periods of time. Even vibrational healing methods will drain a patient if they become over-reliant upon them. It is important to see these as tools for healing and not tools for living. These are the many vibrational tools that are now surfacing into your field of consciousness. Imagine a time when you can rid yourselves of a symptom by listening to music or by placing a fragrance in your field. Such is the future of healing and it has now begun. These vibrational healing techniques will improve in effectiveness as your biology reaches higher vibrational states. Vibrational healing is the modality of the future. Working together with evolving biology will make this global movement much easier.

THE NEW BIOLOGY

Your biology will be altering on many levels. We have mentioned some of the common symptoms that may trouble you during this movement. Let us now tell you of a few of the wondrous things that await you. Your biology was designed to last for hundreds of your years. With the reconnecting of the DNA/RNA, the aging process as you know it, will all but cease. It will be commonplace for you to experience lives of 300 to 400 years. The human biology is fully capable, even today, of complete rejuvenation. This will be realized as humanity moves into this next stage of evolution. Through modalities not yet supported on the Gameboard you will soon find ways of slowing human aging. This has already begun and will be the next area of major breakthrough. As it stands at this moment, these advances will be discovered as a result of searching for cures to the many diseases that inhabit your Gameboard. After the aging process is understood and incorporated, then it will be possible to open the door to age reversal. It is not time to discuss this fully, as there is much still to line up prior to this unfolding. For now, simply know that it is in this direction that you are heading.

As you move further into the changes ahead, you will find communication with your own higher self much easier to accomplish. What you have called channeling will no longer have the mysticism associated with it and will become commonplace. New words will surface to label this natural process of communication. Those words may surprise you. As this happens you will also have clear communications with one another. Up to this point, humanity could not handle being in full contact with one another, as the ego could only operate with separation. This was a necessary illusion caused by polarity on the Gameboard. Being in a survival mode, if the ego had experienced such open communication it simply would have dismissed the experience. This advanced communication will lead to increased awareness of your connection to all things. We have spent much time telling you of the importance of centering your own energy. This was to prepare you for this connection. By centering your own energy, it will be possible to experience full connection to all things without losing yourself in the experience.

There are also many changes occurring in the area of sleep. Many of you in the higher vibrations are already experiencing changing sleep patterns. These will continue to alter as you move forward. These are a natural part of the advancement and are not to be feared. Eventually you will need only brief rest periods to rejuvenate your energy. This will come not only from advanced biology, but also from learning to **channel** energy instead of converting it to heat. More will be forthcoming on this subject. As the Earth continues to raise her vibrations, the pressure on the planet will be felt as a wave encircling the planet. As this wave passes over it will predominantly show as a difficult time to sleep. All on the Gameboard will feel this in some way but those leading the way vibrationally will be more sensitive to it.

LEARNING THE NEW GAME

As you begin to understand the process of aging in human biology, you will also begin to understand the simple concept of what you have termed reincarnation. This opens the door for those of the Crystal Vibration to follow, for they will move rapidly into their work. When this process is understood, even though it will be possible to stay for very long periods, many will choose to leave after a hundred years or so.

The biology is even now shifting at an astounding rate. The DNA/RNA code changes and reconnections are at the very base of this process. There are currently many modalities on the planet that will set this into motion. We tell you that all of these work well. We ask you to find one that resonates with you. Intent is the process that instigates change. These modalities simply provide the stimulus to set your intent into motion. Your choice to step into Plan B has set these changes into motion for you.

THERE ARE NO SUCH THINGS AS ACCIDENTS

We watch your struggles with these changes. We watch as you experience feelings of worthlessness and lose hope. If only you could see yourself the way we see you. The energy of this family of light is colorful and exciting. You are masters that have earned the right to be here at this time. It is not by accident that you are here. It is not by accident that you are reading this now. You have set into motion changes that will alter things far beyond your own Gameboard. We offer you this information as a re-minder of Home and your true heritage. The colors you have already earned will mark your accomplishments for all to see. Still, you choose to move further. Future generations of your lineage will proudly step forward to claim the heritage you are setting into motion this day. If you find yourself lost and cannot re-member who you are then take a moment and connect to others within this family. We also are available to each one of you. If you but ask, we will gladly bring you Greetings from Home. We thank you for asking.

It is with great pride and honor that we re-mind you to treat each other with respect, nurture one another and play well together... the Group.

There are a lot of very exciting things coming to us as a direct result of walking into Plan B. Imagine being in full contact with your higher self. Imagine not having to sleep one third of your life. Imagine an average life expectancy of 900 years. These things are in motion, thanks to the work we have already done. Many of you are responsible for creating the foundation for this work. We wish to re-member to say thank you for a job well done. Let us take that opportunity now. Nice work!

Now... are you ready for the next assignment?

Chapter 17

The Quill of Re-membrance

✳

Scripting your Plan B Contracts

The Quill of
Re-membrance

My plan 'B' contract is

The Quill

*A*s we travel to connect Lightworkers in many places on the Globe, we see some common threads that seem to affect us all. One thing we all have in common is that we are all looking for our Plan B contracts. Plan A gave us very clearly defined roles to play. Now that we have felt the pull, to many of us it seems as if we are feeling the push to walk away from Plan A without knowing where Plan B really is. From the information the Group has given us, we can clearly see that with empowerment also comes responsibility. The second wave of empowerment clearly puts the responsibility of scripting new contracts into our own hands.

No matter where we go in the world, Barbara and I meet Lightworkers who have been holding the door open for us and other channelers. They have planted and nurtured seeds for many years. Some of these people have been in this energy for a very long time and most of them are very excited about the fact that we are using the same words and principles they have been using for a long time. Many of them were carrying this information before it was popular. The challenge is, now that humanity is taking a gradual turn, these forerunners often feel they have lost their sense of direction. It seems unfair to me that they undertook all the hard work of planting the original seeds only to have others come along and take over to reap the harvest. According to the Group, I am now thinking like a human. When I ask these people, none of them seem upset about how things have played out. They are truly happy to see the light going mainstream. The one thing that all of them seem to have in common is that they don't know where to go from here. It's almost as if, having fulfilled their Plan B contracts early, they are now waiting for further instructions.

There are also those that have raised their vibration to a point where they are no longer aligned with the work they have been doing. They feel they must leave and move into something that matches their new vibrational energy. The question is what? I find this to be a common dilemma faced by Lightworkers throughout the world.

The Group describes the Second Wave of Empowerment as an opportunity for us to take responsibility for scripting out own Plan B contracts. But as we are not accustomed to the way this new process works, we often feel stuck. The Group says we are now carrying more of our own power than has been possible at any time before. Because of the steps we are taking into empowerment, our connection to our own higher selves is stronger. The key to further advancement lies in taking responsibility for intentionally writing our next scripts. We have always written our own contracts, but never before from this side of the veil. This message is for all those who, on receiving the wake up call, open their eyes only to ask:

"What now?"

The Group:

Greetings from Home.

This time is very special for us. In these brief moments we offer you the re-membrance of Home. We are very honored that you have asked for this information. It is with the greatest love for you that we present information for moving comfortably into the higher vibrations of the new planet Earth. We offer you this information for your own discernment and empowerment. We ask that you take only that which resonates within your own heart and leave the rest without judgment. We offer you the feelings and vibrations of Home in an effort to re-mind you of your own magnificence. It is our greatest desire to help you to walk fully into your own power re-membering your own connections to the whole.

What we offer you is nothing more than words on paper. As the words are lifted from the paper and run through your own heart you have an opportunity to apply the information in your own daily lives. This takes the mundane words on the paper and breathes life into them in a way that can be done only on the Gameboard. We have a deep welling of love for you as we see your experience getting easier as a result of finding applications for this information. It is not the information that has the power. The power is within you, as you express your inner truth. The information we offer is only to help you re-member that which has been yours all along. We are deeply honored that you are reaching for this higher truth within yourselves, and we stand ready to assist at every opportunity.

There are many flavors of the truth available in your expression of the Gameboard. We ask you to listen to any that pull at your heart, re-membering to trust your own heart over all others. There is no truth, outside of yourself, that holds more power than that which is within you.

The love we have for you cannot be described in terms that you could understand on a conscious level. Understand that often there are messages that will reach you on many different levels at one time. We understand your confusion at the complexity of some of this information. As we have told you, all truth is simple and uncomplicated. Please understand that we speak to many with different messages through the same words and that is often misinterpreted as complexity. Please do not hang on the individual meaning of each word, but rather, accept the overall message as we speak to your heart with the re-minders of home. There are also times that we speak to your heart directly while your mind is busy with the words. This is as it should be, and is done with the consent of your higher self. As with all things, we ask you to allow only those thoughts that leave you more empowered to reside within your field. Becoming a master of your thoughts is a large step to empowerment.

YOUR SECOND WAKE UP CALL

We take this time together to speak of the next step for those of you moving further into your empowerment. Many of you have felt the wake up call and set your intent to step firmly into Plan B. Some of you have found Plan B contracts waiting for you as you opened the door. Others sit patiently waiting for the doors to open with seemingly nothing happening. Once the intent is set, the energy begins to line up for the co-creation process. If the energy and intent are also not open to receive, then the co-creation will not manifest. To facilitate the move forward into the next contract, it is necessary to fully release all that holds you to the old one. This goes much deeper for you than your first thoughts may take you. Those of you who accepted contracts to distribute the early seeds of light, found it necessary to adopt belief systems that helped you to accomplish the tasks you had chosen. In the lower vibrations of the planet these attitudes and beliefs served you well. The pendulum was moving back toward center but was still very far to one side. These beliefs were the balance to that off-center energy that allowed you to center your energy field and complete your contract. In many cases it is these same beliefs that are now holding you firmly in the old energy.

DEATH OF THE OLD LIGHT, OR THE PHANTOM DEATH

There is a specialized segment of the Master Healers on the Gameboard that we have called the Aboriginal Healers. These are the group of honored Master Healers who have agreed to sow the seeds of light since the beginning, even when it was less than popular. Many of you on the Gameboard now have been holding this energy faithfully, only to find yourselves stagnant at a time when many others are awakening. We tell you that without the work you have done, this awakening would not have been possible. The original Plan A contracts of the vast majority of the Aboriginal Healers is now complete and these beloved beings are now faced with a new choice.

For those who choose to walk forward into new contracts, it is necessary to release all attachments to the old contracts that have been completed. The Game was originally scripted to include the alternate outcomes of Plan A and Plan B. When scripting your Plan A scenario, you included an ending that would have brought disasters and Earth changes that would have removed a large part of the population of the planet at this very time.

We will now talk of an experience that many have had or will encounter. This is the experience known as the **Phantom Death.** This experience marks the time that you originally scripted to leave having completed all of your Plan A contracts. Even though you have now altered your direction, the original energy imprint of your proposed exit from the Gameboard may still manifest as a Phantom Death experience. This phenomenon enables your spirit to fully release your attachments to old contracts so that you can begin anew.

This explains why many of you are going through drastic changes in many areas of your lives. Following the Phantom Death experience people will change quite dramatically. This affects relationships in all areas because it alters who you are.

Not everyone will experience the Phantom Death in the same way. Some will experience this in dream state although it will **not** be a dream. Others may experience it as a traumatic event or health challenge. Even though these may be quite severe they will not be long in duration. Some may not re-member the experience at all, even though it will have a profound effect on their energy. As always on the Gameboard of Free Choice, it is your option to return Home during this experience. Even though the pull of Home is very strong very few will take this course as they have spent so much energy positioning for what is now at hand. This is a true experience in every detail; some may even see the Greeters who contracted to meet them upon transition. This experience is one that will gift you with new beginnings. We ask you to embrace this opportunity to re-evaluate the beliefs that have so effectively brought you to this point.   View the Game before you with the wide, eager eyes of a child, for that is very close to the reality from our perspective.

SCRIPTING WITH THE QUILL

The scenario for the Gameboard, as you originally wrote it, allowed for the possibility of Plan B only if the overall vibration of the Earth were raised beyond a certain level.   Even as the first few began to move into the process you have termed ascension, it was not thought that many on the Gameboard would actually awaken. You have shifted this outcome many times over.

We tell you here that because of your choices, some of the portals of energy that were not originally intended to open until the year 2012 are now being made available to you.   The time has come. The pull to move into your purpose is undeniable. This is your chance to take responsibility and begin to create your own reality.   It is time to make decisions about your next steps and to set your intent.   Take your power and do so purposely or it will be done for you by default.   The energy and the Planets are now aligned; opportunities are present for the next stage of work for those who choose to take the Quill and script them.

Previously, all contracts of higher vibration were scripted on this side of the veil.   Know that never before during the Game have you had the power to script your contracts while in a bubble of biology. Yet here you are, standing at the door of your greatest potential. Even though the veil keeps you from seeing who you really are, your magnificence is evident as the God within each of you once again shines through.   Release the beliefs that bind you to the lower vibrations and seek the next step with passion.   Look only to joy in life, for the God within you deserves nothing less.   This joy marks the highest potential for your next steps on the Gameboard.   We have *spoken of the crystals that have marked your path thus far. These carry the vibrations of home and have served to mark your path of least resistance. Now, many of you have reached the part of the path where the crystals appear to end. Fear not, for your heart contains the connection to your higher self that is not restricted by the veil. Release your judgments, set your intent and have the courage to follow your joy. This will lead you straight to the creation of Home on your side of the veil.*

"PLAN B" JOB DESCRIPTIONS

"What is my Plan B contract?" This is the main question of Lightworkers. We ask you to shift your focus ever so slightly from **searching** to **creating,** for this is your true power. If only you knew how powerful you truly are. Your higher self is now waiting for the next move from the Gameboard. You have earned the ability to carry your full power while still in the bubble of biology. All that remains is for you to exercise that power.

While it is not possible to show you your own Plan B contract, it is possible to show you some possibilities. What we will share with you is a view from a perspective of energy. In your humanness you identify readily with labels and descriptions, so we will indulge you here. Please understand that what follows is a suggestion of where to look for your joy and passion. Often these descriptions will overlap and intermingle with one another. We ask that you keep in mind these are words that are our representation of energy as viewed through the eyes of the Keeper of the Sword. As these roles develop in your experience, they may carry other labels. You may find that you incorporate several of these attributes. This is not a definitive listing, but rather an opportunity to see some of the potentials from a higher perspective. What we are doing here is to complicate this just enough for you to understand. Enjoy the reflection as you view yourself in these higher vibrational potentials.

Master Healers

TRANSITION TEAMS
VIBRATIONAL HEALERS
ABORIGINAL HEALERS

A Master Healer is one who has mastered the art of Healing. In many instances, this mastery has been accomplished over several lifetimes and carried forward. **A Healer is one who creates space for others to feel safe enough to heal themselves.**

This can be accomplished in a variety of ways and modalities on the Gameboard. Many of you within this family are of this vibration. The vast majority came in holding this energy quietly, for there were very few asking for healing in the lower vibrations of the Game. At this time there are many awakening Master Healers, for with the evolution of humanity, there is much to be done.

Transition Teams are a special segment of the Master Healers that focus on easing the transition. The work of this segment of Healers will have a vast impact on humanity as a whole in times to come. These are distinguished Healers indeed.

Vibrational Healers are the honored division of the Master Healers that hold the keys to advancing biology. All within your field is expressed as vibration. These higher modalities will offer tools for moving and living comfortably in the higher vibrations of the New Planet Earth.

Aboriginal (Original) Healers are the ones who are largely responsible for planting the seeds that have led to the advancement of humanity. These are the beloved beings that, even in the face of adversity, have faithfully grounded the energy of Light. Most of them will readily create new roles for themselves, once again placing themselves at the cutting edge of advancement.

[For further information about Healers please see the Chapter on "Awakening the Master Healers."]

Architects

ARCHITECTS OF THE HEART
ARCHITECTS OF THE LIGHT

Architects are the designers of the systems that will lead you to full reunion with your power. These are the beloved beings that will make room on the Gameboard for the empowered human. These are the ones who will devise systems of governments and communities that will make it possible to live in harmony and peace, and also to make space for the return of the Children of Crystal Vibration.

Architects of the Heart are also usually Master Healers. These are gentle, powerful beings that find great joy and passion in balancing the heart center of the whole of humanity. This is usually accomplished through the implementation of projects and systems that incorporate the balanced male/female energies on the Gameboard.

Architects of the Light are a special group of beings who will develop systems that will enable everyone to incorporate the higher aspects of Light into daily life. As humanity returns to ethereal biology, this will provide a pivotal role. Much will be needed in new systems and paradigms as humanity moves to higher states. These people will be at the forefront of designing and adjusting these systems to support higher vibrational being. Very few are on the Gameboard at this time. Many will emerge as the vibrations increase.

Cosmic Connectors

NETLINKERS

WEAVERS OF THE NET, OR "LATTICEWORKERS"

Cosmic Connectors are those seen now on the Gameboard who bring people of like vibration together. The majority of these beings are also Aboriginal Healers. They naturally carry a vortex of energy that comfortably attracts others of similar vibration into their field. An important role they are playing in your time is to provide space for the re-union of spiritual family. These grand beings revel in connecting fragmented energy strands. This creates more places for the Light to travel, and therefore creates more Light for all.

Netlinkers are a special faction of the Cosmic Connectors who find expression through the use of technology. Often blending spirituality, biology and technology, they encourage advancement in all these areas. Many of these are now finding ways to work with technology even though they are highly spiritual beings. These will be the very first to understand the link between technology and spirituality.

Weavers of the Net, or "Laticeworkers", are of a very high vibration. They are not yet on the Gameboard, but the forerunners are now in place, to plant seeds for their work yet to come. These are Universal energy workers that will move the Game to a fuller understanding of the connection to the Universal energy grid. As you advance to higher vibrational states, the Universal grid also raises its vibratory rate. The connection is inseparable. These Laticeworkers will learn and develop techniques to energetically work with the Universal Energy Grid and make these adjustments.

Energy Workers

GUARDIANS OF THE UNIVERSAL ENERGY
INTEGRATORS OF LOVE (BLENDED ENERGY)

Energy Workers are a group of people who are experts at observing and working in the natural flow of Universal Energy. Mapping the flow and proper positioning within this flow are their main focus. By placing intent in the flow at just the right location all things can be accomplished effortlessly. These are sought-after teachers and advisors in many areas of life on the new Planet Earth.

Guardians of the Universal Energy are a select few that work in very high levels of organizations and governments. They oversee the actions of these organizations to ensure they are in accord with the Universal energy flow as it relates to the evolving Gameboard. The natural flow of energy with the Universe is to find balance through blending. All actions in opposition to this flow, especially actions of the collective, produce ripples in this flow and restrict the energy available to all. These players will be responsible for viewing all actions of the collective, and comparing the flow of these actions to the flow of Universal Energy. Like all leaders in the higher vibrations, they are trusted, dedicated servants who work to empower.

Integrators of Love are people who are constantly searching for ways to ground the Universal Energy in our daily lives. Many of these will be leaders and educators of the highest order. They will offer humanity methods for integrating the highest form of Light into daily life on the Gameboard.

Mechanics

Mechanics are those who have a talent for "doing." Currently, their greatest joy is to find expression of their creative talents through the use of their hands. As humankind evolves, their work will move to higher and more critical levels. These are the honored doers who bring completion to the energy circle. Unlike the time you now occupy, these workers will be in these positions by choice and not by default. Without these beings all would come to an abrupt halt. In your time these beings work together with architects, providing technology to the Gameboard.

Librarians of Light

KEEPERS OF THE AKASHIC RECORDS
MENTORS OF KNOWLEDGE (EDUCATORS)
KEEPERS OF THE FLAME

Librarians are the ones who will eventually take their place as the Directors of Knowledge. These are people who love to help others find answers. Many are Master Healers who have found ways of healing by leading others to find their own truth. These are the people who will play a large role in helping others to find and use their own power. Much the way librarians of your time lead others to find their truth by leading them to the information in the library, these grand beings will hold vast information and be every ready to direct people to appropriate places to find the knowledge they seek. Soon, many will want to know how it was done on the Gameboard of Free Choice. These are the people who will point the way to that information. A loving, tolerant demeanor is an important attribute that these grand entities carry.

Keepers of the Akashic are the historians who accurately keep the records so that mis-directions of energy need not be repeated. Human history has a tendency to repeat itself if not constantly reminded of lessons learned. These souls will proudly provide that reminder service in a most empowering and loving way. These are much like the historians in your time. The role will become even more important as the course of events unfolds. With the forward movement greatly increased, it will be easier to forget the great lessons you have gained. These beings will provide a valuable base for your advancement.

Mentors of the Knowledge are educators who work side by side with others to integrate the highest of lessons into daily life. Seen in your time as teachers, these are people who find great expression in helping others to locate and learn the tools to move forward. Most of these revered beings also hold the title of Master Healer and this is the way they express their healership.

Keepers of the Flame are a very select group of specialists. They are highly esteemed Mentors that focus entirely on the Children of the New Earth. The challenges that face the children of the New Planet Earth have never before been encountered. These dear beings work entirely planting the seeds of your new Game. These are the ones who now work with the Indigo children who are rearranging the paradigms of humanity. Soon they will hold the responsibility to create space for the return of the Children of Crystal Vibration. They are revered and trusted servants, faithfully holding the flame for all of humanity.

Guardians of the Earth

GROUNDERS OF THE ENERGY
KEEPERS OF THE GRID

Guardians of the Earth are the trusted servants that have accepted the responsibility to hold the sacred energy of the Mother. Their work began recently, when the holders of the energy turned over to humans the guardianship of the Earth. The Whales and Dolphins have been entrusted with the safeguarding of the energy of the planet but that is now humanity is of a high enough vibration to hold that responsibility once again. The cetaceans may stay on as stewards in service to humans, but now, humanity has advanced enough to firmly hold the responsibility of the care of the Mother energy. These beings will stand at the forefront of this cause and proudly stand to serve as examples of how to work in full accord with the Gaia energy.

Grounders of the Energy are a fun-loving group of beings that firmly ground the energy of the higher vibrations into your existence in the third, and soon the fifth, dimensions. They relate to the pleasures of the Game. These beings have been seen in the past as those having hedonistic tendencies. We tell you now the work they do in grounding and balancing the energy has made possible much of your advancement. In your time you have perceived these people as being of lower vibration. We tell you that they do work on levels that are very important to the balance of the Gameboard, and this will be a respected vocation indeed.

Keepers of the Grid are beings dedicated to maintaining the grids. Their work will begin in the area of discovering the physical grids that encircle the planet and provide habitable energy space. Once these discoveries are revealed these people will spring into action to create conscious interfaces with these grids. As the understanding of the grids of the planet increase, these beings will begin to work with the Universal Energy Grid, thus becoming Latticeworkers.

Ambassadors of Light

Ambassadors of the Light are the ones who carry the Light to others. They have been seen as leaders, teachers, shamans, gurus and many other labels that you have used. They are all Master Healers and many also carry the attributes of Mentors. These are dearly loved beings, for they have selflessly placed themselves in a position of speaking their truth in all situations. They have set the energy at the very high personal risk of falling prey to their own ego. As the motivation of humanity moves from survival to achieving unity, their task will remain similar in nature, but they will be scrutinized even more closely as discernment becomes more widely practiced. This process will move them from Teacher to Facilitator, and, eventually, to that of Validator. Together with the Architects, they will devise new methods to speak their truth.

Propagators of the Seeds

HARVESTERS OF LOVE

Propagators of the Seeds are beings that will work together with the Earth to provide a comfortable environment. Some of their activities are like those of a Farmer in your time. They work in conjunction with the Guardians of the Earth, providing products from the essence of the Earth. When this work is combined with the conscious awareness that the Mother is a living, breathing entity, then this work will move to the next level and be utilized to its fullest. Much will be learned about the interaction with the Mother, and this will lead to purposely planting seeds of energy and Light.

Harvesters of Love work energetically with the Propagators of the Seeds, yet their role is quite different. These gentle beings are the ones who will harvest the natural energy of the Earth. There are many natural energy resources that lay untouched in your time. In planting the seeds of energy and Light these beings will find the ways of harvesting these products of the Earth.

Scientists and Alchemists

Scientists and Alchemists of your future will enable the blending of the physical and metaphysical sciences. This will open many doors for the Master Healers who work with biology. As the physical bodies continue shifting to carry more light, it will prompt the search for an even higher truth. These scientists will connect the energy strands in this area. The incorporation and study of energy will open doors not yet imagined. The study of the very small and the study of the very large will lead to the discovery that even chaos holds order. These truths will lead you to a re-membering of your own power and your true position within the Universe.

What we have described to you here is an energy view of typical Plan B contracts. This list is certainly not complete, and we ask you to be patient if the attributes of your heart's work were not covered here. These labels and titles are offered only to suggest to you roles that may be a part of your higher existence. Some of these are not yet supported on the planet and this alone is one reason many feel as if they are not moving forward. In this case, the collective vibration of humanity is not yet supporting the work you will be doing. In these instances we ask you to focus on the attributes that call you to this work, and to find ways to use those attributes in other areas of your daily lives. This will provide you with the joy that you seek and bring you the vibrations of Home.

For us to bring you this information has brought us great pleasure. The Love we feel for you brings us completion beyond your understanding. We too are an integral part of the whole of God and, in that manner we are true family. This message particularly has been of great joy to us. By offering this information in this fashion, we too have had the opportunity to play the Grand Game of Hide and Seek through your eyes. For this we are truly great-full. [sic] Our connection to you is one that runs very deeply. Recently our *connection has grown much stronger to you. We tell you that it is not we that have moved. Your advancement has opened many doors as the collective vibration of humanity is now in the process of building Home on your side of the veil. It is in the deepest respect and highest of honor for you and your work that we ask you to treat each other with respect, nurture one another and play well together... the Group*

You may have noticed the slightly different "flavor" of this channel. Most of the time the Group speaks in a collective voice that is easily recognizable. They began in familiar voice, but after a few paragraphs, just prior to the job descriptions, the energy shifted a bit. They told me the entity that had spoken was someone that we would regard as the Architect of the Group. It's also interesting to note that this entity seemed to have a slightly different reference to time throughout this message.

When I first began doing seminars based on the information from the Group, they told me that I would be given three tools to use in the three dimensions. The Sword I already had. The Scepter I was given shortly after I started the work. The third tool remained a mystery for two years. Recently, I was awakened in the middle of the night as they showed me the third tool. Both Barbara and I are often awakened in the middle of the night and we are quite used to it, usually not giving it another thought. This time, however, was different. Although Barbara did not see me in the dark, she felt the importance of what had happened and asked out loud . . . "What is it?" I was just returning from wherever they had taken me so I didn't want to answer at first. Then she asked again, a little more forceful this time... "What?" I answered, "They just gave me the third tool." At that moment I realized where this conversation was heading and did not want to get into a three-hour discussion at three a.m. Then, before I could open my mouth again, she asked the question: "What is it?" I responded with "It's a Quill." "That's nice," she said, as she rolled over, returning to dreamland. About twenty minutes later I was again awakened from my sleep. In a voice that was obviously awake for some time Barbara asked: "What's a Quill?" I couldn't tell if the laughter I heard was from me or from the Group.

The Quill of re-membrance is used to ground the energy. We use it to script our next contracts and set them into motion by setting our intent in writing.

Until this time the energy was not right for the third tool to be used. Now, because of the advancements we have made, the time is right. It is also no co-incidence that two days after writing this information, we were scheduled to do the first Quill seminar. I had written the seminar based on what they showed me that night in bed. After receiving this information, I went back and re-wrote the entire two- day workshop. In Sudbury, Ontario we held the first Quill of Re-membrance seminar where we actually walked through the Phantom Death, released and cleared our energy, and then wrote our new scripts for Plan B. It is truly the dawning of a new Light on the planet. Along with this new power also comes the responsibility of taking the Quill and actually scripting our future. Now this is being supported. It seems that once again the Master of Time had been holding his finger high in the air, signaling that the time was not yet right for the release of this information. Once again, this Grand Master of Time has just lowered his finger.

Chapter 18

The Return of Merlia

The Return of Merlia

THE RETURN OF MERLIA

*T*here is a wonderful silence in the air at this moment. Listen, and you will hear the heartbeat of the Earth Herself. The anticipation is great. Throughout the kingdom there is awareness that the magical energy is about to re-visit the Earth. This energy has visited the Earth before in other forms to aid at crucial junctures. We know this truth by many names. There is one rule that must be observed prior to any visitation on the Earth. We, the players on the Gameboard, must intentionally ask for it.

There have been many times in our past here on the Gameboard when we did ask for this energy to visit. In most instances, we were not ready to accept the information it offered. Most often we refused to accept its truth. Still, it always came when we asked. The energy came willingly to plant seeds of truth that oftentimes would not be reaped until generations later. With each return it offered an abundance of unconditional love and compassion. This energy most often manifested as a teacher, for this was our way of understanding in the lower vibrations of planet Earth. Now, for the first time this energy prepares to return to Earth as an equal energy that walks by our side. As humans begin to carry more of their own power, we can now see that walking side by side with this energy will quickly lead us to the creation of Home on this side of the veil.

I have attempted to write about this several times, and every time there has been a block to putting this information out in its rightful form. I am once again feeling the gentle nudge from the Group to offer this story of a grand tale. Somehow, this time I think it will be presented.

The rest of the Group falls silent in reverence as She steps into the room. Here, in all of Her magnificence, She stands before us, ready to walk alongside each one. Due to the work we have done and the clearing we have accomplished, we have called Her in. The Time of Merlia is upon us.

The Group:

Greetings from Home.

We have greeted this gathering of masters many times with a heartfelt honoring. We have attempted to show you with each one of our visits just how special you are to us and to all that is. We have had only limited success in conveying to you how much you are loved for the experiences you endure on the Gameboard. You look at your daily life and see only as far as the end of your nose. You do not see that each obstacle you overcome opens the door for many others. We have seen this clearly, and we thank you for your work. You have successfully increased the vibratory rate of the collective. You have taken control at a critical time on the Gameboard. You have accepted your power when it was handed to you, even though you did not fully understand it. The colors you will carry from this time forward will mark you among the elite of the Universe. You are the special ones, for you are the players of Free Choice. The work you have done has made the planet safe for many to follow. This day we speak to you of the return to the Gameboard of the Merlia energy. This is only possible because of the Lightwork you have done for so long. The clearing of yourselves for the purpose of car-rying Light, and the clearing of the planet, has made room for this return. We are honored to play a role in the introduction of this material, for it will forever change the manner in which you play the Game.

THE DAYS OF CAMELOT: AN INTERSECTION OF REALITIES

Many in your higher vibrations have felt the undeniable pull of the tales you call Camelot. You may know that these are myths that have been written for your amusement. No matter the image held in your head, you also know in your heart that they hold great truth. We will offer a simple explanation, for this is so often how spirit speaks to each one of you. We have explained how your own advancement on the DNA level has opened your perceptions of other dimensions of time. We tell you that these stories have found their way into your reality as carriers of truth that began at the intersection where these dimensions cross one another. As a path of reality travels forward following the duality of linear time, it often crosses the path of another reality occupying the same time and space. These realities intersect for only a brief moment, yet as they move on, they leave an indelible impression. The energy of each reality leaves its stamp on the other, leaving behind permanent memories lasting for eons. It is no wonder to us that these incidents are often recorded. Because the realities differ so much, they are most often written about as myths or fictional stories. We ask you to follow the truth of the information here, and not your judgment of its origin.

The stories of these times have been placed in your reality for a reason. Look at the many important seeds planted by this one intersection of alternate realities. The story of the boy Arthur who would be King with no royal lineage behind him, speaks of the right to claim the true power that each one of you holds. The nature of the Game you are playing is to re-member your powers, and to use those powers to re-create Home on your side of the veil. Was it not King The idea was proven that to expect greatness from one another would set the stage for that greatness to surface. The Arthur who created his idea of Heaven on Earth under the name of Camelot? It was striving for the attainment of this goal that set the stage for so many others to create their own ideas of Heaven on Earth. This led to a space in the new government where room was created for the empowered human. The concept portrayed by the round table was that no one person was to be in charge. This speaks clearly of the principles of individual empowerment that we have been offering you. Allowing space for all to take their own power is the way of emulating Universal energy. The overall paradigm has now shifted from follow the leader to follow your own truth, as space is now made for the empowered human. In actuality this is also the return of a form of government that you found very successful in the times of Lemuria. This is actually a system of non-government that will be brought back by the Architects of the Light as they begin to move into their new roles.

In the past, each time this energy visited the planet it was by contract. These contracts were set up to provide the Gameboard with an opportunity to reach the highest conclusion of the Game and create Home on your side of the veil. As the dimensions crossed each other, seeds were planted. If the vibrations of Humanity were high enough, the seeds would sprout and take root. In times past this was not supported because the overall vibration of humanity, and thus the Earth, were not high enough to support these seeds. Therefore, each time this energy visited the Earth; it was not supported and left only its impression as a myth behind as a re-minder. From our view this is seen as an impression in the energy time line. In your world it is a story that is passed from generation to generation. Now the energy returns for the last time. This time the vibrations of the Game are moving rapidly to support the energy. Due to your advancements, this is the time that the creation of Home on Planet Earth is possible. You have made it so.

YOUR TRUE POWERS OF CREATION

Now we will speak again of your true power. We tell you time and again that you have no idea how powerful you really are. You are a finite expression of the infinite Creator and as such you have the same powers as the Creator. You have only to re-member that they are present. Where you allow your thoughts to reside creates your own reality in each moment. This is confusing for you because of the time lag that has served you well as a buffer. The time lag has protected you well from your own destructive thoughts. This was necessary to build in the Gameboard because humans were not yet masters of their thoughts and to have you hold he full power of co-creation was to invite you to manifest your worst fears. In the higher dimensions to which you are aspiring you will see this much more clearly, for the time lag is lessening even in your world at this moment. The power you hold is magical indeed, and more power-ful even than you can imagine. As the veil thins, the time lag will continue to lessen and you will see more readily how quickly you manifest your own thoughts.

Merlin, the Master of Time

We equate this to the time of Camelot when the figure of power was referred to as Merlin. Merlin was a craftsman of power and its uses. Human as he was, he had studied and learned the ways of the magician. We tell you this is much like the powers you are now uncovering within yourselves. Like Merlin, it is necessary for each one to exercise these muscles for full development as an apprentice. When these powers are fully understood the deeds of Merlin will seem pale by comparison. Let us also re-mind you that Merlin was the one who lived backward in time. His linear time frame was always emerging from the future. This is a new concept for most of you for you are not aware that you have control over time. This idea of intentional time control illustrates similar concepts we have helped you re-member. Merlin was the representation of natural power, because his power was equal to his alignment with the

Universal principles. His strength came from his union with all that is. Such is true for all power. No one can be powerful solely unto themselves, for it is the union with all things that generates the power.

There are many seeds that were successfully planted in the collective consciousness of the planet with this single illustration. The hope that was presented in the symbol of Excalibur stands to this day. This is the same sword that represents the angelic purpose of the family of Michael. This is the Sword of Truth and always stands to re-mind you that your power is in standing firm with what you know in your own heart. This is the Sword of Michael and the stands to re-mind you to stand firm in your truth to be fully in your power. In the days of Camelot it was only the boy of pure heart who was able to remove the sword from the stone. The integrity of the boy Arthur was an illustration of the truth as known from within. Finding that truth within, and having the courage to follow that truth, is what makes the connection stronger. Exercising this connection is the process of learning to walk with full connection to your higher self. Find your sword and hold it dear.

Until recently, the illustrations represented in these stories were usually seen through the eyes of the male energy. This has to do with polarity on the Gameboard. Polarity has provided you with the means of existence on the Gameboard, yet it taints your vision drastically. You see things as separate when they are not. This leads to a belief in lack when there is none. As you believe, then so you create. This is the simple nature of the Game as played on the Gameboard of Free Choice. Movement into Plan B means moving from a field of polarity to a field of unity. This is a concept that may be difficult for some to assimilate. Some of your belief systems have been built on judgment rather than discernment. It is these judgments that create energy strings that always tie you to that which you have judged. We re-mind you that we have often said that there is no judgment on our side of the veil, other than that which you bring with you. Judgment is only possible within a field of polarity. As you move out of this field of polarity it will be necessary to re-evaluate these core belief systems. Opening to the potential of unity brings new strength and wisdom. This is now in progress on your planet. You are now in movement from a field of polarity to a field of unity. With that movement, comes a new complete vision.

It is this movement that has cleared the way for the full expression of female energy to return to planet Earth. This is not a time of dominance or competition, for those are strictly illusions of polarity. As seen through the eyes of whole beings, this is a return to full and complete natural power. Because of the clearing work you have done on this planet, this is now possible. The full feminine energy can once again return to the planet and make it whole. This is the full complement of the Yin and the Yang supporting each other in their unity. This is a return to the whole reflection of God, as expressed within each one of you.

THE TIME OF MERLIA

It is with great pride that we tell you that the time of Merlia is now upon you. It is you that have made this possible through your choices on the Gameboard. Because of the work you have chosen to do, you have made it possible for the full feminine energy to return and unite with the whole. You can clearly see that this reunion has begun on your planet. The numbers on your planet have never been so large and yet your crime rates have continued to decline. You are making a difference and changing your world with each thought that enters your hearts. It is this balance between the head and the heart that so clearly shows the re-membering of Merlia. Making space for this part of each one of you will enable the re-membering process to continue. Balance the third dimensional *thinking with the feelings from your own heart and watch the magic return.*

Merlia is the Merlin energy returning to Earth in feminine form to offer the balance needed for the step out of polarity into wholeness. This return has been prophesied in many of your writings. We ask you to make space for Merlia as she returns to offer her balance. She does not come to sit in judgment or to be a leader of humanity. This was indicative of the old energy. The Second Wave has opened a new energy of walking within self-empowerment. She offers to walk alongside each one of you and balance the energy of the planet naturally. She represents an opportunity to utilize your full power by incorporating all aspects of yourself. Until recently, it was not safe for this energy to return, but through your choices you have made it so. You now see evidence of this progression in your collective thinking. Women are making advances in many areas on the Gameboard at this time. This will continue as this energy gently assimilates and naturally finds a balance. This energy has been mistrusted and suppressed for some time. Now, it will find its way back to seek a natural balance. Balancing the heart and the head will open doors that will lead to great potential.

Once the equilibrium of energy is reestablished on the planet, the doors will open for the next phase of the evolution of humanity. This will be the introduction of those of crystal vibration onto the planet. These are gentle beings that will be the next incarnation of humanity on planet Earth. They will help you to understand the full implications of walking hand in hand with your own higher self. This will mark a time of peace on the planet that has never before been known.

You Have Cleared the Path for the Return

It is your own daily struggle with life in the third dimension that has cleared the way for this to unfold. It was only possible to build the Gameboard in this fashion. Many of you look at your daily life and feel stuck because there is so much you want to do. So often you say, "If only I had this, or that, then I would be able to do what I came here to do." We tell you that with the veils firmly in place it is not possible for you to see the effect you have already had. We tell you that you have already changed the outcome many times over. It was only possible in this manner. You have chosen to redefine the Game. These seeds of your evolution and the new Game have already been planted. You have done very well preparing the soil. We are deeply honored to be a small part of this process. Through your actions and choices you have called the original families back together once again. This reunion is a grand one indeed and is not limited to your side of the veil. We, too, have been reunited with a part of our family that we thought was all but lost. We are joyful to see you return. Welcome Home.

It is with the greatest of love that we ask you to treat each other with respect, nurture one another and play well together... the Group

The birds sit in silence as the wind holds its breath. Anticipation is everywhere. A miracle is about to happen.

Final Chapter

The Game Continues...

The Grand Game of Hide and Seek
Part II

On the Gameboard in Holland

The Game Continues

Opening your eyes as if from a long sleep, you find yourself in a crowd, moving as part of the whole. There is a charge of excitement in the air as the crowd moves toward its destination. We all turn the corner and find ourselves in the exact location at the base of the mountain where this Game originally began. Each one takes the same seat they had the day we first designed the Gameboard. A call has gone out to all places for the players to gather once again. As you settle in, look around to see all the members of your family of light. The memories of home instantly fill you with a joy that you had almost forgotten.

Ahhh! to lift the veil for even a moment brings such joy. The vibrations of Home flood your being, and in that instant you are whole and free. This body is much lighter than the one you are used to. As this feeling emanates within your being, you re-member the times on the Gameboard when you experienced these glimpses of Home. All the pieces are beginning to fit together again. Now it is all coming back to you. "Wow! What a Game! I can't believe those veils work so well!" You hear yourself say.

You look to your sister standing next to you and recognize her as one you know from the Gameboard. For a moment a wave of anger overcomes you as you identify her as one that played the part of an enemy. She smiles, and in that instant your memories trail further back and you see her as one carrying a deep love for you. The wonderful vibrations of Home return once again as these feelings of anger leave you. Then you re-member a planning session where you asked this loved one to play the enemy. She was very hesitant, but in the end she agreed to play the part because she loved you so. Then your clarity returns and you re-member that it was only a Game. With the clarity also comes a strong knowing that you are a part of the whole. It's funny how that works. It seems that the two vibrations of love and fear cannot occupy the same space at the same time.

A brother now stands and calls this family meeting to order. There is a reason we are here for this brief interlude from the Game. We are here to discuss the options that lay before us. In fact, someone is here to present those options to us. She is one you all know deep within your heart. She was with us from the very beginning, and now She is here to help us balance for this next phase of the Game. As the family grows silent, she is introduced with a reverence that reflects her prominence.

"It is with the greatest of love that I give you~ ~ ~ ~ ~ ~ Merlia."

Complete silence falls over the crowd and Merlia speaks:

"The information I bring with me this day is accompanied with the greatest of honor from all sides of the veil. You on the Gameboard have played your parts far beyond what was expected, and what I bring you today are crystals of honor to carry with you always. These are given to you now in honor of your part on this Gameboard of Free Choice. You have taken this great experiment to an unexpected level, a level that has allowed us to alter the paradigm of all that is to come. The work you have done here has reached far beyond the Gameboard we originally devised. As a reminder, we offer these crystals to carry with you always. These crystals may be used in many ways in your experience on the Gameboard. Doors will open in your path, for these crystals identify you as one from this great family of light. With this crystal your vibrations will precede you in all things. All that encounter you from this day forward will know of your magnificence and the role you played on the Gameboard of Free Choice. Your name will be known as one that forever opened the door for all to follow."

Just then, she waves her hand and a tingling feeling vibrates throughout your whole being. Looking down you see a beautiful crystal lying in your hand.

She continues speaking, and with each word you feel the connection to her and everything around you.

"You have played the grand Game and brought it to its highest conclusion. As this grand Game draws to a close the plan set forth is now in need of alteration. As the last moments of the Game approach, many of you have begun to re-member your powers. It is therefore time to redefine the Gameboard of Free Choice to make room for the expression of this power.

"Many of you have been on the Gameboard time and again, only to find how efficiently the veils were designed. Many times have we passed each other along our paths, only to gaze into one another's eyes and not recognize who we are. It has been a grand experiment, and much was gained from your willingness to walk behind such effective veils. On the Gameboard of Free Choice you have looked into your hearts and re-membered the part of the whole that you are. This was not expected, but it has allowed a new understanding of the one truth. Your reflection of the infinite was grander than expected.

"As the Gameboard shifts to the next level you will have the option to stay and play in the new energy if you wish. For those choosing to stay, it will be a time of change as you raise your vibrational level to match that of the new Earth and those of crystal vibration. Your biology will continue to adjust as a new vibration presents itself on the Earth. These changes may seem challenging for some, as you have no reference for them. Please turn to each other during this process, as it will ease the strain. This alteration will allow you to comfortably carry your own light. It will allow you to step into the next level of the Game while still within your bubble of biology.

"A time of peace will now reign on the new planet Earth. This is a peace that was earned by the toil and effort of you who

purposely directed the Light on the Gameboard. You have suc-
ceeded far beyond our wildest dreams. You have made it safe for
those of Crystal Vibration to join you on the Gameboard. This is a
time of great joy and the fulfillment of a destiny that is long overdue.
This is a time of peace and harmony. This is a time when the love of
light will reign on the Earth. We have purposely chosen this future
for the Gameboard. We have chosen well.

"We will step into this future by re-membering our heritage.
There is one easy way to set into motion the changes you desire.
The easiest way to accomplish this task is to look at yourself through
the eyes of those belonging to your original spiritual family. This is
why we have brought you all back together at the location where
you originated the Game. Now is a time when you have called
these parts of yourself back into your field on the Gameboard as
well.

"Learn the art of walking side by side with one other. Nurture
one another with unconditional love as you did in the first days of
the Game. Re-member the original plan and make it so. You are
connecting to family to allow you to re-member your true nature and
your power. When we come together in this fashion, we gain
strength from one other. Even brief encounters forever alter your
experience. Make space in your lives to find these people and look
into their eyes. Here we will see the truth and re-member who we
really are.

"The family of light to which you belong is one of greatness.
Your heritage is a lineage of master healers who have walked the
Gameboard. Seek them out and welcome them into your field with
open arms. Play in the energy the way you did when the Game was
new. Help all to re-member who they are by looking through one
another's eyes.

"Once this truth has visited you it will not be possible to fall back into lower vibrations. Send out the call and seek them out, for they hold the truth about you and your true family of light.

"You have played the Grand Game and brought it to its highest conclusion. You, the purveyors of light, have made it so. Step boldly into a future of your own design. Stay if you choose, as we rebuild the garden. In your travels, always carry the story of how it was done. This is your new heritage and it is a grand one indeed. These are the colors you have earned and they will be with you always. Walk purposely into your own passion and fearlessly create Home on your side of the veil with the power that is rightfully yours. It is time to take this power and carry it proudly. With every step, re-member your new heritage and always hold your head high, for you are one of this great family of light. It is time to fully re-member."

As the last echo of her voice faded she transformed into a dragonfly and was gone. Somehow, you know you'll see her again, and in that knowing a sense of completeness fills your being. Rising now to leave, you re-member the words of Merlia - about making room for your original spiritual family. Then you look to a person standing next to you. For just a second that person returns your glance. In that moment your brain is filled with memories that flood your senses. These memories are from a long forgotten time. A magical connection to this person makes everything seem possible. Moving from one person to the next, now, as each one crosses your field of vision, you find an opportunity to look into their eyes and re-member. Here the truth can hide no longer. Now it is moving.

It has begun.

A sense of gratitude fills your being as you understand how special it is to be here at this time. The next level of the Game holds so many possibilities. It will be wonderful to walk in daily life and hold this much energy. This will be a time of greatness on the planet and you will be very much a part of it. Returning home now, you find yourself sitting in the chair where you began this journey. A question forms in your mind. "What if I lose my direction? What if I can't re-member?" Then your hand begins to vibrate in a way that has only happened once before. Looking down you see a beautiful crystal sparkling in the palm of your hand,

...a smile fills your heart,

...and you **Re-member**.

Connect with spiritual family:

http:// www.Lightworker.com

Re-minders from Home can be delivered monthly via e-mail by sign-ing up at the web site for the Beacons of Light.

More information about the 𝔓aths to 𝔈mpowerment 𝔖eminars, which are based on the writings from the Group, is available at:

http://www.Lightworker.com/schedule.shtml

The Beacons of Light Internet Meditations and monthly re-minders from Home are available by request at:

http://www.Lightworker.com/beacons/snailmailjoin.shtml

The writings and monthly messages from the Group can be found at:

http:// www.Lightworker.com/beacons/

Find others on the message boards and in the chat rooms.

Re-member... You are not alone...

The on-line magazine that presents many "flavors of the truth."

http://www.PLANETLightworker.com

Built for Lightworkers by Lightworkers. This is a magazine format web site that provides information from the cutting edge of the higher vibration.

Regular Sections include:

Children of the New Earth

Vibrational Healing

Writings in the Light

Beyond the Veil

The Angelic Realm

Regular monthly columns written by leading authors and inspirational writers in the metaphysical/spiritual arena.

Come Join us!

Paths to Empowerment Seminars

Paths to Empowerment Seminars provide practical applications of the information for living in the higher vibrations of the new planet Earth, based on information from the Group. All gatherings include practical techniques for healers and channeling exercises to help you open to channeling your own information, together with a Live channel from the Group through Steve Rother. The Following three Seminars were designed to implement the information in this book.

The Lightworker Spiritual Re-union

Based on the Sword of Michael, this seminar is centered around applying the first three tools of the higher vibrations: Co-Creation, Discernment and Synchronicity.

The Scepter of Self Love

Provides practical tools, such as meditating in an elevator, centering your energy in the midst of chaos, creating your life in 26 seconds and time warping.

The Quill of Re-membrance

Works with the Art of Graceful Acceptance,

Shifting Dimensional Realities, Relationships in the New Energy and Scripting your Plan B contracts.

Check the schedule on the web site and watch for new seminars and special gatherings. The Paths to Empowerment Seminars from Lightworker are listed at:

http://www.Lightworker.com/schedule.shtml

You will receive notification of events in your area by adding your name to our mailing list at:

http://www.Lightworker.com/beacons/snailmailjoin.shtml

Lightworker sm Is a non-profit corporation dedicated to spreading
Empowerment and Light.

Bringing Home to this Side of the Veil

We at Lightworker have a dream. If you feel drawn to this work and would like to be part of the team on any level we invite you to write, call, or visit our web site and request information on how you may help.

Lightworker
P.O. Box 1496
Poway, Ca. 92074-1496
(858) 748 5837

Information is also available on the web site at:

http://www.lightworker.com/helpwanted/

Join the Lightworker Mailing List

Send your Name, Street Address, Phone and E-mail to:

Lightworker
Department 555
P.O. Box 1496
Poway, Ca 92074-1496

Or add your information on-line at:
http://www.lightworker.com/beacons/snailmailjoin.shtml

About the Author

Steve Rother was comfortably settled into life as a General Building Contractor in the San Diego area when, through an odd series of events, he was placed firmly in the middle of his contract. Steve and his wife Barbara began shifting their focus on life and living on New Year's morning, 1995, when they found themselves unexpectedly expressing their intent for the coming year during a ceremony that took place as the sun rose over a California beach. From that day forward, their lives were never to be the same.

Soon after, Steve began channeling "the Group" in writings that subsequently became the Beacons of Light Monthly Re-minders from Home. These monthly Internet meditations, which started on America Online in 1996 with just a handful of people, are now distributed to thousands of people in 112 countries at press time. These monthly writings with messages from "the Group" are about re-membering and accepting our own power, and living comfortably in the higher vibrations now on planet Earth. The Group calls Steve the "Keeper of the Sword". The Sword of Truth is the sword of Archangel Michael and is being used today for personal empowerment so that we never have to use it as a weapon again.

Steve never returned to his contracting business. Today, he and Barbara, his wife of 28 years, travel the world, presenting seminars of personal empowerment to Healers, Lightworkers and enlightened Corporations based on information from the Group. They have presented these in many countries, and have twice been invited to speak at the United Nations in Vienna, Austria. On the second of these visits in April of 2000, Steve presented a class on channeling and spirit communication, believed to be the first such class ever presented at the UN.

"Our purpose is to help people find their own power and inner guidance. When we Re-Member who we are, then we can go about the task of Creating Heaven here on Earth, one heart at a time. It is this act that will lead us into the higher vibrations and our next stage of evolution as spirits in human form."

Steve and Barbara live in San Diego, California, where they work together in love and Lightwork. They have formed the non-profit corporation of Lightworker and, together with volunteers and staff, work at planting seeds of personal empowerment on a global basis. More information about Steve, Barbara and the Group, including their seminar schedule, can be found at the web site: http://www.lightworker.com

Barbara & Steve Rother in the Vienna Woods

Index

$\mathcal{R}e$-member - ORDER FORM

1. Online with your Credit Card via a secure server at:
   http://www.Lightworker.com/bookstore/
2. Fax your orders on the form below to Lightworker
   at (858) 748 7640
3. Telephone orders Toll Free at (877) 248 5837
   Outside the US call 01 858 748 5837
4. Order by Mail by sending this order form with check payable
   to Lightworker or Credit Card info to:

Lightworker
PO Box 1496
Poway Ca. 92074-1496

Please send _____ copies of **Re-member $14.95 each**

Name _____

Address _____

Address 2 _____

City _____State _____ZIP _____

Telephone _____

E-mail _____

Please add sales tax of 7.75% if delivered inside California.
Shipping: US: $4 first book, $2 each additional book
International: $6 first book and $3 each additional copy
All Prices in US Funds

We accept Mastercard, Visa, American Express and Discover

Card Number _____Exp _____

Name on Card_____

Signature _____

$\mathcal{R}$e-member **- ORDER FORM**

1. Online with your Credit Card via a secure server at:
   http://www.Lightworker.com/bookstore/
2. Fax your orders on the form below to Lightworker
   at (858) 748 7640
3. Telephone orders Toll Free at (877) 248 5837
   Outside the US call 01 858 748 5837
4. Order by Mail by sending this order form with check payable
   to Lightworker or Credit Card info to:

Lightworker
PO Box 1496
Poway Ca. 92074-1496

Please send _____ copies of **Re-member $14.95 each**

Name _____

Address _____

Address 2 _____

City _____State _____ZIP _____

Telephone _____

E-mail _____

Please add sales tax of 7.75% if delivered inside California.
Shipping: US: $4 first book, $2 each additional book
International: $6 first book and $3 each additional copy
All Prices in US Funds

We accept Mastercard, Visa, American Express and Discover

Card Number _____Exp _____

Name on Card_____

Signature _____

Re-member - **ORDER FORM**

1. Online with your Credit Card via a secure server at:
   http://www.Lightworker.com/bookstore/
2. Fax your orders on the form below to Lightworker
   at (858) 748 7640
3. Telephone orders Toll Free at (877) 248 5837
   Outside the US call 01 858 748 5837
4. Order by Mail by sending this order form with check payable
   to Lightworker or Credit Card info to:

Lightworker
PO Box 1496
Poway Ca. 92074-1496

Please send _____ copies of **Re-member $14.95 each**

Name_____

Address _____

Address 2 _____

City _____State _____ZIP _____

Telephone _____

E-mail _____

Please add sales tax of 7.75% if delivered inside California.
Shipping: US: $4 first book, $2 each additional book
International: $6 first book and $3 each additional copy
All Prices in US Funds

We accept Mastercard, Visa, American Express and Discover

Card Number _____Exp _____

Name on Card_____

Signature _____